AGRICULTURAL PLANNING AND SOCIAL JUSTICE IN INDIA

AGRICULTURAL PLANNING AND SOCIAL JUSTICE IN INDIA

Manish Dubey

CYBER TECH PUBLICATIONS
4264/3, Ansari Road, Daryaganj, New Delhi-110002 (India)
Ph.: 011-23244078. Fax: 011-23280028
E-Mail: cyberpublicationsdelhi@yahoo.com

AGRICULTURAL PLANNING AND SOCIAL JUSTICE IN INDIA

Manish Dubey

First Edition 2009

Published by :

G.S. Rawat for **Cyber Tech Publications**
4264/3, Ansari Road, Daryaganj, New Delhi-110002 (India)
Ph.: 011-23244078. Fax: 011-23280028
E-Mail: cyberpublicationsdelhi@yahoo.com

Printed at :

Preface

Although agriculture contributes only 21% of India's GDP, its importance in the country's economic, social, and political fabric goes well beyond this indicator. The rural areas are still home to some 72 percent of the India's 1.1 billion people, a large number of whom are poor. Most of the rural poor depend on rain-fed agriculture and fragile forests for their livelihoods.

The sharp rise in food-grain production during India's Green Revolution of the 1970s enabled the country to achieve self-sufficiency in food-grains and stave off the threat of famine. Agricultural intensification in the 1970s to 1980s saw an increased demand for rural labor that raised rural wages and, together with declining food prices, reduced rural poverty.

Sustained, although much slower, agricultural growth in the 1990s reduced rural poverty to 26.3 percent by 1999/00. Since then, however, the slowdown in agricultural growth has become a major cause for concern. India's rice yields are one-third of China's and about half of those in Vietnam and Indonesia. With the exception of sugarcane, potato and tea, the same is true for most other agricultural commodities.

The Government of India places high priority on reducing poverty by raising agricultural productivity. However, bold action from policymakers will be required to shift away from the existing subsidy-based regime that is no longer sustainable, to build a solid foundation for a highly productive, internationally competitive, and diversified agricultural sector.

This book is written specially in this regard to serve the purpose of the readers and the people who are somehow engaged with agriculture or any type of agricultural activity.

—Author

CONTENTS

1

History of Agriculture

Agriculture was developed at least 10,000 years ago, and it has undergone significant developments since the time of the earliest cultivation. Evidence points to the Fertile Crescent of the Middle East as the site of the earliest planned sowing and harvesting of plants that had previously been gathered in the wild. Independent development of agriculture occurred in northern and southern China, Africa's Sahel, New Guinea and several regions of the Americas. Agricultural practices such as irrigation, crop rotation, fertilizers, and pesticides were developed long ago but have made great strides in the past century. The Haber-Bosch method for synthesizing ammonium nitrate represented a major breakthrough and allowed crop yields to overcome previous constraints. In the past century agriculture has been characterized by enhanced productivity, the substitution of labor for synthetic fertilizers and pesticides, selective breeding, mechanization, water pollution, and farm subsidies. In recent years there has been a backlash against the external environmental effects of conventional agriculture, resulting in the organic movement.

ORIGINS

Identifying the exact origin of agriculture remains problematic because the transition from hunter-gatherer societies began thousands of years before the invention of writing. Nonetheless, archaeobotanists/paleoethnobotanists have traced the selection and cultivation of specific food plant characteristics, such as a semi-tough rachis and larger seeds, to just after the Younger Dryas (about 9,500 BC) in the early Holocene in the Levant region of the Fertile Crescent. There is earlier evidence for use of wild cereals: anthropological and archaeological evidence from sites across Southwest Asia and North Africa indicate use of wild grain (e.g., from the ca. 20,000 BC site of Ohalo II in Israel, many Natufian sites in the Levant and from sites along the Nile in the 10th millennium BC). There is even evidence of planned cultivation and trait selection: grains of rye with domestic traits have been recovered from Epi-Palaeolithic (10,000+ BC) contexts at Abu Hureyra in Syria, but this appears to be a localised phenomenon resulting from cultivation of stands of wild rye, rather than a definitive step towards domestication.

It isn't until after 9,500 BC that the eight so-called founder crops of agriculture appear: first emmer and einkorn wheat, then hulled barley, peas, lentils, bitter vetch, chick peas and flax. These eight crops occur more or less simultaneously on PPNB sites in the Levant, although the consensus is that wheat was the first to be sown and harvested on a significant scale.

Mehrgarh, one of the most important Neolithic (7000 BC to 3200 BC) sites in archaeology, lies on the "Kachi plain of Baluchistan, Pakistan, and is one of the earliest sites with evidence of farming (wheat and barley) and herding (cattle, sheep and goats) in South Asia."

By 7000 BC, sowing and harvesting reached Mesopotamia and there, in the super fertile soil just north of the Persian Gulf, Sumerian ingenuity systematized it and scaled it up. By 6000 BC farming was entrenched on the banks of the Nile River. About this time, agriculture was developed independently in the Far East, probably in China, with rice rather than wheat as the primary crop.

Maize was first domesticated, probably from teosinte, in the Americas around 3000-2700 BC, though there is some archaeological evidence of a much older development. The potato, the tomato, the pepper, squash, several varieties of bean, and several other plants were also developed in the New World, as was quite extensive terracing of steep hillsides in much of Andean South America. Agriculture was also independently developed on the island of New Guinea.

In China, rice and millet were domesticated by 8000 BC, followed by the beans mung, soy and azuki. In the Sahel region of Africa local rice and sorghum were domestic by 5000 BC. Local crops were domesticated independently in West Africa and possibly in New Guinea and Ethiopia. Evidence of the presence of wheat and some legumes in the 6th millennium BC have been found in the Indus Valley. Oranges were cultivated in the same millennium. The crops grown in the valley around 4000 BC were typically wheat, peas, sesame seed, barley, dates and mangoes. By 3500 BC cotton growing and cotton textiles were quite advanced in the valley. By 3000 BC farming of rice had started. Other monsoon crops of importance of the time was cane sugar. By 2500 BC, rice was an important component of the staple diet in Mohenjodaro near the Arabian Sea. By this time the Indians had large cities with well-stocked granaries. Three regions of the Americas independently domesticated corn, squashes, potato and sunflowers.

THEORY

The reasons for the development of farming may have included climate change, but possibly there were also social reasons (e.g., accumulation of food surplus for competitive gift-giving as in the Pacific Northwest potlatch culture). Most likely there was a gradual transition from hunter-gatherer to agricultural economies after a lengthy period during which some crops were deliberately planted and other foods were gathered in the wild. Although localised climate change is the favoured explanation for the origins of agriculture in the Levant, the fact that farming was 'invented' at least three times elsewhere, and possibly more, suggests that social reasons may have been instrumental.

When major climate change took place after the last ice age c.11,000 BC much of the earth became subject to long dry seasons. These conditions favoured annual plants which die off in the long dry season, leaving a dormant seed or tuber. These plants tended to put more energy into producing seeds than into woody growth. An abundance of readily storable wild grains and pulses enabled hunter-gatherers in some areas to form the first settled villages at this time. There are several theories as to what drove populations to take up agriculture:

- The Oasis Theory which was original proposed by Raphael Pumpelly in 1908, but popularized by Vere Gordon Childe in 1928 and summarised in his book *Man Makes Himself* This theory maintains that as the climate got drier, communities contracted to oases where they were forced into close association with animals which were then domesticated together with planting of seeds. It has little support now as the climate data for the time does not support the theory.
- The Hilly Flanks hypothesis. Proposed by Robert Braidwood in 1948, it suggests that agriculture began in the hilly flanks of the Taurus and Zagros mountains, and that it developed from intensive focused grain gathering in the region.
- The Feasting model by Bryan Hayden suggests that agriculture was driven by ostentatious displays of power, such as throwing feasts to exert dominance. This required assembling large quantities of food which drove agricultural technology.
- The Demographic theories proposed by Carl Sauer and adapted by Lewis Binford and Kent Flannery. This leads from an increasingly sedentary population, expanding up to the carrying capacity of the local environment, and requiring more food than can be gathered. Various social and economic factors help drive the need for food.
- The evolutionary/intentionality theory. As proposed by those such as David Rindos the idea that agriculture is an evolutionary adaptation of plants and humans. Starting with domestication by protection of wild plants, followed specialisation of location and then domestication.

ANCIENT AGRICULTURE

By the Bronze Age, wild food contributed a nutritionally insignificant component to the usual diet.

If the operative definition of *agriculture* includes large scale intensive cultivation of land, mono-cropping, organized irrigation, and use of a specialized labour force, the title "inventors of agriculture" would fall to the Sumerians, starting ca. 5,500 BC. Intensive farming allows a much greater density of population than can be supported by hunting and gathering, and allows for the accumulation of excess product for off-season use, or to sell/barter.

The ability of farmers to feed large numbers of people whose activities have nothing to do with agriculture was the crucial factor in the rise of standing armies. Sumerian agriculture supported a substantial territorial expansion which along with internecine conflict between cities, made them the first empire builders. Not long after, the Egyptians, powered by farming in the fertile Nile valley, achieved a population density from which enough warriors could be drawn for a territorial expansion more than tripling the Sumerian empire in area.

SUMERIAN AGRICULTURE

In Sumer, barley was the main crop, but wheat, flax, dates, apples, plums, and grapes were grown as well. Mesopotamia was blessed with flooding from the Tigris and Euphrates rivers but floods came in late spring or early summer from snow melting from the Turkish mountains.

With Salt deposits under the soil, all of this made Mesopotamia very hard to farm. The earliest known sheep and goats were also domesticated and were in a much larger quantity than cattle.

Sheep were mainly kept for meat and milk, and butter and cheese were made from the latter. Ur, a large town that covered about 50 acres (20 hectares), had 10,000 animals kept in sheepfolds and stables and 3,000 slaughtered every year. The city's population of 6,000 included a labour force of 2,500 cultivated 3,000 acres (12 km²) of land. The labour force contained storehouse recorders, work foremen, overseers, and harvest supervisors to supplement labourers. Agricultural produce was given to temple personnel, important people in the community, and small farmers.

The land was plowed by teams of oxen pulling light unwheeled plows and grain was harvested with sickles in the spring. Wagons had solid wheels covered by leather tires kept in position by copper nails and were drawn by oxen and the Syrian onager (now extinct). Animals were harnessed by collars, yokes, and headstalls.

They were controlled by reins, and a ring through the nose or upper lip and a strap under the jaw. As many as four animals could pull a wagon at one time. Though some hypothesize that Domestication of the horse occurred as early as 4000 BC in the Ukraine, the horse was definitely in use by the Sumerians around 2000 BC.

CHINESE AGRICULTURE

The unique tradition of Chinese agriculture has been traced to the pre-historic Xianrendong Relics and Diaotonghuan Relics (c. 12 0000 BC-7500 BC). Chinese historical and governmental records of the Warring States (481 BC-221 BC), Qin Dynasty (221 BC-207 BC), and Han Dynasty (202 BC-220 AD) eras allude to the use of complex agricultural practices, such as a nationwide granary system and widespread use of sericulture.

However, the oldest extant Chinese book on agriculture is the *Chimin Yaoshu* of 535 AD, written by Jia Sixia. Although much of the literature of the time was elaborate, flowery, and allusive, Jia's writing style was very straightforward and lucid, a literary approach to agriculture that later Chinese agronomists after Jia would follow, such as Wang Zhen and his groundbreaking *Nong Shu* of 1313 AD.

Jia's book was also incredibly long, with over one hundred thousand written Chinese characters, and quoted 160 other Chinese books that were written previously (but no longer survive). The contents of Jia's 6th century book include sections on land preparation, seeding, cultivation, orchard management, forestry, and animal husbandry. The book also includes peripherally related content covering trade and culinary uses for crops.

For agricultural purposes, the Chinese had innovated the hydraulic-powered trip hammer by the 1st century BC. Although it found other purposes, its main function to pound, decorticate, and polish grain that otherwise would have been done manually.

The Chinese also innovated the square-pallet chain pump by the 1st century AD, powered by a waterwheel or an oxen pulling a on a system of mechanical wheels. Although the chain pump found use in public works of providing water for urban and palatial pipe systems, it was used largely to lift water from a lower to higher elevation in filling irrigation canals and channels for farmland.

Papuan Agriculture

Ancient Papuans are thought to have begun practicing agriculture around 7000 BC. They began domesticating sugarcane and root crops. Pigs may also have been domesticated around this time. By 3000 BC, Papuan agriculture was characterised by water control for irrigation.

Indian Agriculture

Evidence of the presence of wheat and some legumes in the 6th millennium BC have been found in the Indus Valley. Oranges were cultivated in the same millennium. The crops grown in the valley around 4000 BC were typically wheat, peas, sesame seed, barley, dates and mangoes.

By 3500 BC cotton growing and cotton textiles were quite advanced in the valley. By 3000 BC farming of rice had started. Other monsoon crops of importance of the time was cane sugar. By 2500 BC, rice was an important component of the staple diet in Mohenjodaro near the Arabian Sea.

The Indus Plain had rich alluvial deposits which came down the Indus River in annual floods. This helped sustain farming that formed basis of the Indus Valley Civilization at Harappa. The people built dams and drainage systems for the crops.

By 2000 BC tea, bananas and apples were being cultivated in India. There was coconut trade with East Africa in 200 BC. By 500 AD, eggplants were being cultivated.

Roman Agriculture

Roman agriculture built off techniques pioneered by the Sumerians, with a specific emphasis on the cultivation of crops for trade and export. Romans laid the groundwork for the manorial economic system, involving serfdom, which flourished in the Middle Ages.

Mesoamerican Agriculture

In Mesoamerica, the Aztecs were some of the most innovative farmers of the ancient world and farming provided the entire basis of their economy.

The land around Lake Texcoco was fertile but not large enough to produce the amount of food needed for the population of their expanding empire. The Aztecs developed irrigation systems, formed terraced hillsides, and fertilized their soil. However, their greatest agricultural technique was the *chinampa* or artificial islands also known as "floating gardens".

These were used to make the swampy areas around the lake suitable for farming. To make chinampas, canals were dug through the marshy islands and shores, then mud was heaped on huge mats made of woven reeds. The mats were anchored by tying them to posts driven into the lake bed and then planting trees at their corners that took root and secured the artificial islands permanently. The Aztecs grew corn, squash, vegetables, and flowers on chinampas.

Andean Agriculture

The Andean civilizations were predominantly agricultural societies; the Incas took advantage of the ground, conquering the adversities like the Andean area and the inclemencies of the weather. The adaptation of agricultural technologies that already were used previously, allowed the Incas to organize the production a diversity of products of the coast, mountain and jungle, so them could be able to redistribute to villages that did not have access to other regions.

The technological achievements reached to agricultural level, had not been possible without the workforce that was at the disposal of the Sapa Inca, as well as the road system that was allowing to store adequately the harvested resources and to distribute them for all the territory.

MUSLIM AGRICULTURAL REVOLUTION

From the 8th century, the medieval Islamic world witnessed a fundamental transformation in agriculture known as the "Muslim Agricultural Revolution", "Arab Agricultural Revolution", or "Green Revolution".

Due to the global economy established by Muslim traders across the Old World during the "Afro-Asiatic age of discovery" or "Pax Islamica", this enabled the diffusion of many crops, plants and farming techniques between different parts of the Islamic world, as well as the adaptation of crops, plants and techniques from beyond the Islamic world, distributed throughout Islamic lands which normally would not be able to grow these crops.

These techniques included crop rotation, irrigation and pest control. Some have referred to the diffusion of numerous crops during this period as the "Globalisation of Crops",which, along with increased mechanization of agriculture, led to major changes in economy, population distribution, vegetation cover, agricultural production and income, population levels, urban growth, the distribution of the labour force, linked industries, cooking and diet, clothing, and numerous other aspects of life in the Islamic world.

Serfdom became widespread in eastern Europe in the Middle Ages. Medieval Europe owed much of its development to advances made in Islamic areas, which flourished culturally and materially while Europe and other Roman and Byzantine administered lands entered an extended period of social and economic stagnation. As early as the ninth century, an essentially modern agricultural system became central to economic life and organization in the Arab caliphates, replacing the largely export driven Roman model.

The great cities of the Near East, North Africa and Moorish Spain were supported by elaborate agricultural systems which included extensive irrigation based on knowledge of hydraulic and hydrostatic principles, some of which were continued from Roman times.

In later centuries, Persian Muslims began to function as a conduit, transmitting cultural elements, including advanced agricultural techniques, into Turkic lands and western India. The Muslims introduced what was to become an agricultural revolution based on four key areas:

- Development of a sophisticated system of irrigation using machines such as norias, water mills, water raising machines, dams and reservoirs. With such technology they managed to greatly expand the exploitable land area.
- The adoption of a scientific approach to farming enabled them to improve farming techniques derived from the collection and collation of relevant information throughout the whole of the known world. Farming manuals were produced in every corner of the Muslim world detailing where, when and how to plant and grow various crops. Advanced scientific techniques allowed leaders like Ibn al-Baytar to introduce new crops and breeds and strains of livestock into areas where they were previously unknown.
- Incentives based on a new approach to land ownership and labourers' rights, combining the recognition of private ownership and the rewarding of cultivators with a harvest share commensurate with their efforts. Their counterparts in Europe struggled under a feudal system in which they were almost slaves (serfs) with little hope of improving their lot by hard work.

- The introduction of new crops transforming private farming into a new global industry exported everywhere, including Europe, where farming was mostly restricted to wheat strains obtained much earlier via central Asia. Spain received what she in turn transmitted to the rest of Europe; many agricultural and fruit-growing processes, together with many new plants, fruit and vegetables. These new crops included sugar cane, rice, citrus fruit, apricots, cotton, artichokes, aubergines, and saffron. Others, previously known, were further developed. Muslims also brought to that country lemons, oranges, cotton, almonds, figs and sub-tropical crops such as bananas and sugar cane. Several were later exported from Spanish coastal areas to the Spanish colonies in the New World. Also transmitted via Muslim influence, a silk industry flourished, flax was cultivated and linen exported, and esparto grass, which grew wild in the more arid parts, was collected and turned into various articles.

Agriculture in the Middle Ages

During the Middle Ages, Muslim farmers in North Africa and the Near East developed and disseminated agricultural technologies including irrigation systems based on hydraulic and hydrostatic principles, the use of machines such as norias, and the use of water raising machines, dams, and reservoirs.

They also wrote location-specific farming manuals, and were instrumental in the wider adoption of crops including sugar cane, rice, citrus fruit, apricots, cotton, artichokes, aubergines, and saffron. Muslims also brought lemons, oranges, cotton, almonds, figs and sub-tropical crops such as bananas to Spain.

Renaissance Agriculture

The invention of a three field system of crop rotation during the Middle Ages, and the importation of the Chinese-invented moldboard plow, vastly improved agricultural efficiency.

After 1492 the world's agricultural patterns were shuffled in the widespread exchange of plants and animals known as the Columbian Exchange. Crops and animals that were previously only known in the Old World were now transplanted to the New and vice versa.

Perhaps most notably, the tomato became a favorite in European cuisine, and maize and potatoes were widely adopted. Other transplanted crops include pineapple, cocoa, and tobacco. In the other direction, several wheat strains quickly took to western hemisphere soils and became a dietary staple even for native North, Central and South Americans.

Agriculture was a key element in the Atlantic slave trade, Triangular trade, and the expansion by European powers into the Americas. In the expanding Plantation economy, large plantations producing crops including sugar, cotton, and indigo, were heavily dependent upon slave labor.

British Agricultural Revolution

Between the 16th century and the mid-19th century, Great Britain saw a massive increase in agricultural productivity and net output.

New agricultural practices like enclosure, mechanization, four-field crop rotation and selective breeding enabled an unprecedented population growth, freeing up a significant percentage of the workforce, and thereby helped drive the Industrial Revolution.

By the early 1800s, agricultural practices, particularly careful selection of hardy strains and cultivars, had so improved that yield per land unit was many times that seen in the Middle Ages and before.

The 18th and 19th century also saw the development of glasshouses, or greenhouses, initially for the protection and cultivation of exotic plants imported to Europe and North America from the tropics.

Experiments on Plant Hybridization in the late 1800s yielded advances in the understanding of plant genetics, and subsequently, the development of hybrid crops..

Increasing dependence upon monoculture crops lead to famines and food shortages, most notably the Irish Potato Famine (1845–1849).

Storage silos and grain elevators appeared in the 19th century.

Recent History

New Technologies

With the rapid rise of mechanization in the late 19th and 20th centuries, particularly in the form of the tractor, farming tasks could be done with a speed and on a scale previously impossible.

These advances, joined to science-driven innovations in methods and resources, have led to efficiencies enabling certain modern farms in the United States, Argentina, Israel, Germany and a few other nations to output volumes of high quality produce per land unit at what may be the practical limit.

The development of rail and highway networks and the increasing use of container shipping and refrigeration in developed nations have also been essential to the growth of mechanized agriculture, allowing for the economical long distance shipping of produce.

While chemical fertilizer and pesticide have existed since the 19th century, their use grew significantly in the early twentieth century.

In the 1960s, the Green Revolution applied western advances in fertilizer and pesticide use to farms worldwide, with varying success.

Other applications of scientific research since 1950 in agriculture include gene manipulation, and Hydroponics.

New Criticisms

Though the intensive farming practices pioneered and extended in recent history generally led to increased outputs, they have also led to the destruction of farmland, most notably in the dust bowl area of the United States following World War I.

As global population increases, agriculture continues to replace natural ecosystems with monoculture crops.

In the past few decades, western consumers have become increasingly aware of, and in some cases critical of, widely used intensive agriculture practices, contributing to a rise in popularity of organic farming, the growth of the Slow Food movement, and an ongoing discussion surrounding the potential for sustainable agriculture.

2

LIBERALISATION, THE STATE AND AGRICULTURE IN INDIA

In the early 1980s, the Indian state initiated the process of liberalisation although the formal announcement was made in 1991. We know that the Indian economy, unlike in China or in Eastern Europe, was never fully regulated and hence reforms have been relatively mild and smooth.

(1) In India the agricultural sector rests in the hands of few capitalist farmers who thrive at the expense of small and marginal farmers.

(2) The majority of the subsidies offered by the state were many times useful for only these small sections of large rural agricultural community. The result is that over the years few farmers acquired wealth while majority of them still remain poor.

In India, the reform process has not encountered much resistance from pressure groups because it has not made much headway in respect of debureaucratising and depoliticising the management of the infrastructure and in decentralizing rural development. Nevertheless, India, as one of the founding members of the World Trade Organisation (WTO) has an obligation to open the economy for more liberal international trade and investment by removing restrictive practices in a phased manner. The process of liberalisation and globalisation of Indian economy is likely to influence the production patterns as well as production relations having far reaching implications for the working class in general particularly those operating in the agricultural sector, The ongoing process of liberalisation broadly involves two sets of policies:

1. Macro-Economic Stabilisation — which intends to promote internal and external balances through reduction in public expenditures and subsidies, which influences domestic demand directly, and indirectly by devaluation of currency, monetary policy, etc.; and,
2. Structural Adjustment Policy — which intends to promote economic efficiency through deregulation and integration of Indian economy with the world. This involves a shift in the production structure from non-tradables and replacement of import substitution industrialisation with export orientation.

Hitherto initiated reforms in the national policies relating to domestic industry and foreign investment, industrial licensing, trade, taxation, subsidies, etc., are likely to affect agriculture through spread or spillover effects. But, specific reforms like (i) reduction in subsidies on food, fertilizers, power and irrigation, (ii) liberalization of agricultural trade involving withdrawal of quantitative restrictions on import, decanalisation of agricultural exports and imports, reduction in tariff rates, removal of zonal restriction on agricultural commodity movements, (iii) relaxation of land ceiling and tenancy laws, (iv) removal of reservation of production of certain commodities by the small and household sectors and (v) rural credit reforms involving deregulation of interest rates of rural co-operative banks and relaxation of service area restrictions for bank operations etc., would influence the income and employment of the agriculture sector directly.

Liberalisation in India has far confined itself to industry and trade sectors. The state has not yet come out with a detailed policy package for the agriculture sector and only some ad-hoc measures like raising fertilizer prices have been announced. In the near future, liberalisation policy will be extended to the agriculture sector as well. In the recent budget statement (2000-01) the central government proposes to implement relaxation of land ceiling laws, removal of restriction on production of certain commodities, reduction of subsidies for fertilizers, power, food, etc., which has serious implications for the small and marginal farmers. In this context, the article discusses the impact of liberalisation on Indian agriculture. It also presents the effective means to restructure agriculture in the context of liberalisation.

THE ROLE OF THE STATE IN AGRARIAN TRANSFORMATION

In India, the role of the state in bringing agricultural transformation is made possible by active co-operation between millions of farmers across the country and the governments at the federal and provincial levels. As we observe elsewhere in the text, agriculture in India is policy driven. According to Singh (1995) "the broad policy framework towards agricultural sector in India consisted of the following:

(a) land reforms to remove the intermediary landed interests coupled with measures like consideration of holdings and some half-hearted programs of land redistribution.

(b) the creation of supporting infrastructure in the forms of irrigation facilities, rural roads, electrification, credit and market facilities etc.

(c) creation of an agricultural research, training and extension system on a vast scale, and

(d) a system of public procurement and distribution of agricultural products, particularly food grains with a judicious balance between the interest of the producers and consumers" (Singh, 1995:7).

It is agreed amongst many professional economists (Rao, 1998; Patnaik, 1998; Balakrishnan, 2000; et al.) that agriculture should never be left entirely at the mercy of the free market. State intervention in agricultural markets can be justified on various grounds:

* Unlike industry, where production is a continuous process, agricultural output appears not continuously but at discrete intervals, because of which output cannot

be adjusted to demand conditions with any rapidity.

* The ability of peasants to hold back their produce / stocks is limited in the case of agriculture as the scale of operations is not as big as in industry.
* There is large fluctuation not only in the supply adjustments, but also in the overall output induced by weather and other natural calamities.
* Most importantly, demand for agricultural produce is price-inelastic, which means that agricultural prices and incomes are subject to large fluctuations.

Thus, the responsibility of the state is to ensure that adequate public investments are made in irrigation, agricultural research and infrastructure that ensures well-functioning markets and would facilitate exports. Parikh et al., (1999) point out that "for food security, the government should follow a mixed strategy of modest buffer stocks supplemented by occasional trade on the world market (Parikh et al., 1999:33).

We must note that agriculture is basically a private activity in which public investment has a critical role to play in providing the infrastructure such as roads, electricity, technology, storage facilities, etc.

This not only requires high investments on part of the government but also requires proper direction and focus.

Thamarajakshi (1999) notes that "in this context GCF (Gross Capital Formation) in the public sector in agriculture, both in absolute and relative terms has been rather low whether in the pre-reform or reform period besides recording a decline in the later period; real GCF in agriculture declined by 14.3% between 1985-97 compared to an increase of 13.3% in total public sector GCF while GCF in private sector rose in agriculture, the rise was lower (58%) than that in non-agriculture (74%) apart from the fact that the absolute size of private sector GCF in agriculture, although much higher than public sector GCF in agriculture, was comparably quite low in relation to the size of private sector GCF in non-agriculture" (Thamarajakshi, 1999:2293).

Table: Gross Capital Formation in Public and Private Sectors

(Rs. in crores)

	Agriculture	Non-Agriculture	Total
Public Sector			
1985-91	11347	18229	19576
	(6.9)	(93.1)	(100.00)
1991-97	1155	21020	22175
	(5.2)	(94.8)	(100.00)
Private Sector			
1985-91	2995	19891	22886
	(13.1)	(86.9)	(100.00
1991-97	4736	35195	39931
	(11.9)	(88.1)	(100.00)

Note:(i) Average for each period at constant, (1980-81) prices.;

(ii) Figures in parentheses are percentage total.

Source: Thamarajakshi, 1999:2293.

Thus, in India with the state's emphasis on a market-based system under liberalisation, the overall public investment relating to total investment in agriculture has been constantly declining.

As is evident from Table, there has been a decline in GCF in agriculture vis-a-vis non-agriculture. Also there was a decline both in public as well as private investments. An interesting point to be noted is that private investments in agriculture are more than double than the investments by the public sector. However, during 1993-98, corresponding GCF for public and private sectors was around 6% and 9% respectively. Decline in private sector GCF in agriculture under liberalisation is conspicuous.

According to a report by the Planning Commission (1997), planned investment during 1992-97 (i.e., precisely after completion of Eighth Five-Year Plan) in agriculture was over Rs. 6,49,999 crores, while actual investment was only at Rs.

3,83,000 crores (at 1996-97 prices). The distinction between actual investment to that of planned Public investment in agriculture being 59%(5) (Planning Commission, 1997:12).

Table: GCF in Agriculture (% to total GCF at 1993-94 Prices)

Year	Public Sector	Private Sector	Total
1993-94	6.3	10.7	8.9
1994-95	6.1	10.1	8.9
1995-96	6.4	89.0	8.6
1996-97	6.0	9.4	8.1
1997-98	5.6	8.9	8.3

Note: Percentage in each column is that to total GCF in the respective sectors.

Source: CSO, Revised Estimates of National Accounts.

We know that in many areas of India, irrigation forms the basis for agricultural growth. Investment in this sector is crucial for acceleration of growth in agriculture. According to Table below, total investment in irrigation under liberalisation was around Rs. 5,15,000 crores, which was slightly higher (around 4%) than the earlier period. According to a recent issue of Economic Survey (1999), area under medium and major irrigation increased by 2.1 million hectares during reform period, in comparison to near constancy earlier. However, minor irrigation recorded highest increase of around 14.1 million hectares for the period 1985-97.

Table: Investment in Irrigation (At constant 1996-97 prices)

Power	Major & Medium	Minor	CAD(*)	Total
1985-92	29332.8	1630.2	3684.9	49325.9
1992-97	31057.6	17302.5	3162.9	51523.0

(*) Catchment Area Development.

Source: Planning Commission, Nineth Five-Year Plan.

AGRICULTURAL GROWTH AND PRODUCTIVITY

There is clearly a need for a significant step-up in real state expenditure on agricultural research as well as its optimal distribution over different regions and commodities. This is because an increase in expenditure in agricultural research and extension training leads to increase in overall output of agricultural products. This injection of funds is essential for raising the overall growth rate in agriculture, for realising productivity gains in regions where there is large untapped potential, for reassigning benefits from trade, for alleviation of poverty and for protecting the environment.

In the performance of the 1980s one feature remains unchanged, that is, during the period associated with the economic reforms, the picture of agricultural growth that emerges is as lacklustre as it is widely acknowledged to be. Nevertheless, a slight nuance might be provided. This is that, viewed from a longer term perspective, starting in the 1950s, the economic reforms initiated since 1991 haven't succeeded in reversing the trend in Indian agricultural production which has been downward, except for one phase in the case of the wheat crop: On the other hand, in the 1970s, until the period of the reforms we witnessed a reversal of the buoyancy of the 1980s which has got less attention than is necessary for an appreciation of what drives the growth process in agriculture.

The decline in the rate of growth in the 1990s must be viewed against the background of a significant development in the agricultural sector vis-a-vis the rest of the economy. On the one hand, we are aware that this development which has come as a result of some favourable market changes such as effective price mechanisms for food crops, made during the period by India towards the agriculture sector, alerts us to the likely success of the reforms in a certain sphere. On the other hand, viewed as part of a large picture, it helps us evaluate the explanation of slow agricultural growth that is based on price disincentives. The development we refer to is the dramatic shift in the relative price in favour of agriculture. If slow growth is due to the disincentive to agriculture, allegedly endemic to import-substituting industrialisation, then the reforms since

1991 appear to have succeeded in rectifying the so-called "price distortion."

In India, the rate of growth of agriculture has been higher in the 1980s as compared to 1960s and 1970s. However, there has been a regional concentration of this growth.(6) Sources for agricultural growth in the 1980s include an increase in input use for example the use of fertilisers more than tripled compared to the 1960s. Rao and Hanumappa (1999) say that "the period 1949-50 to 1964-65 could be identified as one of agricultural growth area expansion, the growth rate in area was higher than in yield during this period. The next period, 1967-68 to 1980-81, is usually reckoned the Green Revolution period though it may be appropriate to regard it as the first round of the Green Revolution.

Table: Agricultural Growth, Annual Compound Growth Rate

Period	Area	Yield	Production
1949-50 to 1964-65	1.61	1.50	3.31
1967-68 to 1981-82	0.54	1.83	2.38
1980-81 to 1991-92	0.05	3.16	3.21
1949-50 to 1991-92	0.64	2.05	2.70

Note: Agricultural growth of 1965-66 were excluded as they are years of serious scarcity.

Source: Rao and Hanumappa, 1999:A-133.

The period is marked by sharp reduction in area expansion and a perceptible upward shift in yields... the second round of the Green Revolution 1980-81 to 1991-92, had a more credible record with the almost entire growth rate of 3.21% coming from improvement in yields. The annual growth rate of 3.5% of food grain production during the 1980s was probably due to Green Revolution successes as well as good monsoons during the decade. However, in the 1990s, Indian agriculture could not maintain this pace and annual growth rate fell to 1.7%. Agricultural crop production was about 3.9% in 1998-99 as against a drop of 6% in 1997-98 (Economic Survey, 1998:117).

Table: Annual Growth in Agricultural Production*

Year	Index	% Change
1994-95	165.2	5.0
1995-96	160.7	-2.7
1996-97	175.4	9.1
1997-98	164.9	-6.0
1998-99	171.3	3.9

(*) Index of Agricultural Production (46 principal crops) with base triennium ending 1981-82 = 100.

Source: Economic Survey, 1998.

There was an earlier acceleration in the growth of agriculture despite the fact that the value of modern inputs (fertilisers, pesticides, electricity and diesel oil) at constant 1980-81 prices to the value of gross agricultural output marginally increased from an average 11% during 1985-91 to 13% during 1991-96 and, more particularly, the per hectare consumption of fertilisers rose perceptibly from 56.7 kgs to 70.6 kgs in the periods under comparison.

In fact, the compound growth rate of food grains production dropped to 1.66% per annum during 1990-98 from the peak level of 3.54% per annum during 1980-90, and the average growth rate of gross value of output of agriculture at 1980-81 prices in the reform period was lower at 3.1% per annum than that during the earlier period at 4.3%.

A similar pattern as measured in the recorded growth of gross value of agricultural output has occurred in gross value added.

The average growth rate of GDP from agriculture during 1993-99 was less than that during 1987-99 as well as 1993-99 was lower than that of GDP from the non-agricultural sectors. Since the sectoral distribution of the workforce has remained almost constant in these periods, the differentials in inter-sectoral growth have led to considerably lower levels of per worker value added in agriculture vis-a-vis the non-agricultural sectors. In both periods, the growth rate in the case of the agricultural worker was less than that of his counterparts in non-agriculture.

Table: Per Worker Value Added

	1987-88	1993-94	1998-99
Agriculture	8404	10435	12179
	(4.0)	(3.3)	
Secondary	29519	38684	53276
	(5.2)	(7.5)	
Tertiary	34719	46724	66074
	(5.8)	(8.3)	
All	17022	22160	29353
	(5.0)	(6.5)	

Note: (i) Gross value added at 1993-94 prices derived for 1987-88 on comparable basis.

(ii) Workforce pertains to 15+ age groups.

(iii) Figures in parenthesis refer to average annual rates over the preceding period.

Source: CSO, Revised Estimates on National Accounts at 1993-94 prices.

During 1987-99, the share of agriculture in real GDP declined from 31.6% to 26.8% while the work force remained at 64 to 65%. Consequently per worker value added in agriculture, which was 28 to 24% of that in the secondary and tertiary sectors in 1987-88 declined to 27 to 22% respectively in 1993-94. The relative per worker value added in agriculture vis-a-vis the non-agricultural sectors became more adverse in 1998-99 with the former coming down to 23 to 18% respectively of the respective per worker value added in the secondary and tertiary sectors.

REFORMS IN AGRICULTURE

Policy reforms in the rural sector are both critical and sensitive in nature as well as in their impact because of heavy concentration of working people in rural areas. At present about 40% of total state expenditure in the rural sector is absorbed by direct subsidies, another 22% by anti-poverty programs and only 38% goes to productivity enhancing expenditure, as opposed to 60% in 1981-82.

During the post-reform period, annual growth rate of agricultural subsidies in real terms has substantially declined. But, it remains as high as 3.6% of GDP per annum.

In real terms, the food subsidy increased from Rs. 10.8 billion in 1989-90 to 14.3 billion in 1995-96 at 1980-81 prices, although food subsidy as a percentage of GDP remained constant at about 0.5%.

The revamped Public Distribution System (PDS) has been introduced since 1992, which is targeted to those poor and backward regions where the Employment Assurance Scheme (EAS) is implemented (Annual Report, Ministry of Rural Areas and Employment, 1996). However, Mr. Yashwant Sinha in his Budget, 2000-01 targeted middle and upper classes. From now on, income-tax payers would not get any commodities under PDS, and the others would not get certain commodities such as sugar.

The fertiliser subsidy as percentage of total agriculture GDP declined from 0.9% in 1990-91 to 0.6% in 1995-96. But in absolute real terms, it has been subjected to high annual variations. The fertiliser subsidy policy affected different fertiliser prices differently.

The prices of phosphoric and potassium fertilisers were decontrolled in 1992, but price control on low analysis nitrogenous fertilisers continued. However, in the current year (2000-01) the subsidy price of urea is fixed at Rs. 4,600 per ton. Same for MOP (Potash) is fixed at Rs. 4,260 per ton and for Di-Ammonium Phosphate (DAP) it is Rs. 8,880 per ton. This means that there would be an increase of 15% in prices of urea and MOP and 7% in the case of DAP.

Power and irrigation subsidies, which are supported by the state governments, accounted for nearly 8% respectively of the total agricultural subsidies (as of 1994-95). During the post-reform period (1990-91 to 1994-95), rural power subsidy grew at the rate of 14% per annum in real terms, while the growth rate of irrigation subsidy remained low at 1.2% per annum (based on the Budget Estimates). But irrigation subsidy was reduced mainly on account of non-wage outlays on operation and maintenance and not because of improvement in cost recovery.

Zonal restrictions on the movement of agricultural commodities have been removed since February, 1993, including the lifting of informal controls on wheat movement by private trade.

The excise duty on coffee has been removed. The role of coffee board has diminished and now there is a trend towards open marketing of coffee. Imports of all agricultural commodities other than cereals, oilseeds and edible oils and all agricultural exports (except onion) have been decannalised.

India has agreed to phase out quantitative restrictions on import of 2700 items, out of which 800 are agricultural commodities by April 2003. An agreement had already been reached with European Union and Australia to remove quantitative restrictions on imports from these countries by April, 2000 (Haque, 1997:12).

Agricultural trade reforms initiated in 1991-92 relaxed quantitative restrictions on a few minor commodities. But by 1994-95, it included rice exports and imports of most edible oils, sugar and cotton. Nevertheless, quantitative restrictions on exports of most agricultural commodities except rice continue.

The share of tradable agricultural production protected by non-tariff barriers on the import side, were reduced from about 96% before June, 1991 to 77% in July, 1996. With the liberalisation of sugar imports under 0% tariffs, sugar industry has now to compete with imports.

Similarly, the liberalisation of cotton imports at zero percent tariff has been initiated in 1994. Edible oils are now importable at 20% tariff. The tariff on import of pulses has been reduced from 10 percent to 5 percent. In January, 1997 the U.S. and several other developed countries contended that India no longer suffers from balance of payments problems and therefore, the quantitative restrictions on imports by India would have to be removed immediately.

India made an agreement with European Union and Australia to remove quantitative restrictions on many items in three phases of 3 years, 2 years and 1 year, using April, 1997 as the reference year. But the dispute with U.S. has yet to be resolved.

In the livestock sector, state controls and subsidization of dairy co-operatives continue. The Milk and Milk products Order (1992) presents competition in the dairy industry. The poultry feed manufacturing continues to be reserved for the small scale sector.

The commercialization of fishing has been initiated. Agro-industries that export 50% or more of their output, have been allowed to import their inputs duty free and to import capital equipment at concessional import duty rates w.e.f. April, 1993. In 1991, tractors, combine harvesters and rice transplanters were included in the list of products for which there is now automatic approval of foreign equity of up to 51%.

But several other agricultural implements and farm inputs such as plastic piping, sheeting etc. are reserved for production by small-scale firms.

In the area of rural credit, the reforms include:

(i) reducing target group lending from 100 percent to 40 percent in the case of regional rural banks,

(ii) greater freedom to the banks to rationalize their branches,

(iii) deregulation of interest rates of rural cooperative banks,

(iv) permission to urban co-operative banks to lend to borrowers in continguous rural areas and

(v) relaxation of service area restrictions.

Besides, the Reserve Bank of India (RBI) and National Bank for Agriculture and Rural Development (NABARD) have initiated actions for strengthening Regional Rural Banks (RRBs). Further, the 1996-97 budget provides for doubling the paid up share capital of NABARD and establishing agricultural development financial institutions at the state level to promote investment in horticulture, floriculture and agro-processing.

LIBERALISATION, AGRICULTURE AND THE RURAL POOR

Low productivity and inability to absorb the growing labour force make the agricultural sector in India witness to a pervasive process of marginalisation of the poor inspite of having sustained agricultural growth:

1. There existed policy bias towards large industries and urban areas in contrast to rural areas and agriculture.
2. The agricultural growth policy has benefited limited regions, crops, and farmer groups under green revolution resulting in agricultural capitalism and emergence of haves and have-nots in agricultural sector.
3. The unfavourable policy context for the rural people has resulted in an institutionalized vacuum at the grass root level making it difficult for developmental agencies to reach rural poor and to mobilise them for their growth and development.

As a result of the above, there has been a spiraling effect at the local level thereby inhibiting capacities of the rural people to take development initiatives, provide local leadership and building organizations for collective action to protect their interest. Policy makers choose developmental strategies and priorities that had little concern for these rural populace. The prevailing policy context along with the shift towards liberalization and globalisation has long term implications for human development in India.

It is likely that poverty diminishes with growth — particularly agricultural growth. Liberalisation through its linkage with growth could be a powerful anti-poverty strategy. However, it is not clear that the type of growth that has taken place hitherto in India and which is likely to come in the coming decades would reduce poverty measured in terms of multiple deprivation criteria besides food.

In fact, the marginalisation process in operation in agriculture is very likely to have an extremely adverse impact on poverty measured in terms of low or diminishing human development.

We can broadly characterise Indian Agricultural Policies from 1960- to date in such a way that they focus on promoting capitalist agriculture in the areas undergoing rapid technology change or having promising potential in that direction.

Research, extension, input supply, credit & price and marketing support to help these areas were the primary concern of policy makers. Little did they bother about agrarian reforms and still less in undertaking the massive investments and support measures needed by the vast areas remaining outside the focus of policy makers.

Liberalisation means living the mainstream growth processes fully open to market forces and private enterprise. It is expected that this would lead to a substantial withdrawal of the state from the mainstream economy.

However, the state would have to carry the responsibility for the efficiency and welfare of the peripheries, which the normal profit-oriented economic processes tend to by-pass. In the case of India, an important part of the peripheral area is the vast rainfed area with stagnant agriculture, degraded land resources, underutilised water resources and a substantial number of people barely managing to subsist.

The peripheries need research, extension and thorough going measures for conservation and efficient use of the land and water resources (Rao, 1996:A-61).

The state would have to take a long-term holistic view of these needs and coordinate systematic efforts to meet them. An important step to take is to increase the scope for the Panchayat

Raj Institutions (PRIs) and other participatory organizations to actively contribute to these efforts.

These organisations can also take as large a share as they can in bringing agricultural growth. We know that development of institutions and growing political consciousness at the lower strata as a result of such development (politicisation) would build institutional mechanisms to control, regulate and use of area's resources based on the fair measures of consensus. In such a situation the emphasis of agricultural growth shall be pro-poor.

The emerging situation in India has certain ominous features. The polity remains fragile; the economy is severely marked with strains and suffers from resource crunch, infrastructure and environment. With such a backdrop, promoting growth and maintaining a balance becomes difficult especially in a society, which is ridden with tensions, conflicts and violence as we have in India.

This is more so when promotion of growth is left to the capitalist market forces and the act of balancing is to be ensured by the state itself. The state, as we know, is under constant pressure to do or not to do certain things which inevitably would affect the rural poor.

CONCLUSION AND FUTURE OPTIONS

The benefits of liberalization could not hitherto reach the rural poor because of the failure of economic reforms to penetrate the vital agriculture sector. Indian farmers are potentially no different from other entrepreneurs.

They adopt new technologies enthusiastically, respond to price signals, and grab opportunities to improve their own well being. The principal reforms that are actually needed are to free farmers from all domestic restrictions on storage, transport, and sale of agricultural products.

The present policy of low output prices and subsidized (low) input prices puts a net tax on farmers; we should change to a system of internationally competitive prices for both inputs and outputs. To protect poor consumers,it is much better to provide an effective targeted safety net rather than distort food prices, which has high costs as compared to the benefits.

The announced high priority to agriculture sector remained as mere rhetoric: a concrete policy to guide growth in agriculture was missing. Even the recent Budget does not have any major steps for this sector. This resulted in decline in agricultural production despite good monsoons over the decade. In this context, the next stage of any policy reforms must have the following features:

* More investments to make agriculture boom (especially on developing infrastructure in rural areas such as irrigation, roads, power, etc.). This not only provides livelihood to many, but also becomes vital for industrial development as also for overall performance of the economy.
* Adoption of technology that would suit local conditions. Government should encourage establishing of agro-processing industries in a big way that shall help the farmers to sell finished products rather than selling raw materials.
* Liberalisation of land policy-involving relaxation of land ceiling and tenancy laws, so that, more land can be brought under agriculture. Another alternative is to bring more land under cultivation and promoting viable land holdings through consolidation of holdings held by farmers in fragments. This not only increases productivity but also reduces the use of inputs such as fertilizers and costs for tilling of land by mechanization etc. Consolidation of fragmented holdings would make land conducive for mechanization and judicious mechanization would free the farmers from protracted drudgery.
* State should ensure withdrawal of public sector intervention, except for regulation of production of food grains. Main attention must be paid to reform and restructure the public sector organisations so as to

regulate the agriculture related activities such as transport, water resources / irrigation, etc., with managerial accountability and proper efficiency.

* Phase out restrictions on export of agricultural commodities and services.
* Agricultural production shall continue to be mainly dominated by the private sector. However, the development and management of Common Property Resources (CPRs) such as water, which is main resource for agriculture should be with the state and its actors such as government (central, state, local), non-governmental organisations (NGOs), institutions etc. This action could improve the availability of such CPRs for common use.
* In the areas having irrigation, transport, marketing, processing etc., there shall be private management, which needs proper guidelines and laws. The state should define rules and regulations to check monopoly and ensure fair-trading and usage of CPRs. In this connection, in the rural sphere ensuring active participation of NGOs to monitor and regulate such activities shall bring proper check.
* The effective implementation of the spirit of 73rd Constitutional Amendment by giving the elected local self-government institutions (the three-tier Panchayati Raj Institutions — PRIs at Village, Block and District level respectively) the authority and resources for local development works will make a significant contribution to more effective use of resources. They are better placed to plan works / activities that meet specific needs and possibilities of each panchayat and since periodic elections are mandatory for these bodies, they also create strong inbuilt pressures for public accountability.
* We know that there are umpteen reasons for majority of people depending on agriculture in India. Similarly, there are many follies in India, such as the estimation of workforce depending on agriculture under the census

data. The census data take all those in the rural areas as part of agriculture only. Non-farm activities should also be taken into consideration to get realistic estimate of people depending on agriculture. This apart, there is another possibility. That is, even if whatever workforce that has been absorbed by industry and service sector growth from agriculture sector, it might be nullified by the growth in population. Hence the imperative of population control in the rural areas.

* In order to rectify such lacunae in the analysis of agriculture one has to take a realistic analysis of rural socio-economic conditions. Economic and social empowerment and development of the rural poor has to be achieved basically through productive employment and human development. According to Rao (1998), this requires rapid broad-based, labour-intensive growth. "This implies that India should aim at an annual average growth rate in GDP of 7-8% together with agricultural growth rate of around 4-5% over the next decade."(Rao, 1998:1948).

* Finally, for the successful growth and development of rural India, agriculture growth is a prerequisite. For achieving growth in agriculture, there is need to improve the efficiency of delivery systems for overall rural development that would transform rural life from one of stagnation into a dynamic and vibrant one.

The basic difference in the case of China and Eastern Europe vis-a-vis India is that agriculture was fully regulated in those countries. In China the initiative to reform agriculture has come from the state itself.

The agricultural co-operatives, which were working under communes, were disintegrated and were given to individuals under household responsibility system. Lengthy debates and bitter ideological struggle marked opening of Chinese economy to the outside world. In India, these features were conspicuously missing. Agriculture in India is mostly in private partnership and large farmers / agricultural lobbies benefit at the cost of small and marginal farmers. The naive protests offered by some of the leftists in India to the process of liberalization show ideological bankruptcy. For one reason or another, the rightists as well as parliamentary leftists more or less welcomed liberalization of Indian economy.

The agricultural lobbies of capitalist farmers in India are mostly from North India and West India. They are products of the Green Revolution. There are some farmers' lobbies in the East as well as South, however they do not have as much say as those do from the North India.

It is significant to note that the Budget 2000 speech by Indian Finance Minister Mr. Yashwant Sinha, does not have any maior comprehensive package for the agriculture sector. However, measures mentioned above in the text are to he implemented which for sure affects small farmers, what the government hitherto is pursuing is only to rechannelization / amalgamation of existing schemes for rural agricultural development and poverty alleviation withdrawal of subsidies on food (PDS prices) fertilizers etc., would not only burden the small farmer but become a hurdle in the process of equitable development.

This was also true with the industrial sector as well during the three decades of Five-Year Plan. The protection given to the industries under import substitution has made Indian industry assume monopoly over Indian markets, although the goods produced were not of international standards.

If India enters into world market, prices of majority of the commodities would he affected. Hence, India being a major producer as well as consumer of agricultural products (in majority cases it is one of the leading countries) should enter the world market, so that agriculture exports grow faster than the present. It is time that, India in the context of liberalisation, should think of producing for the world rather than merely on conventional term of consumption purposes.

For example, West Bengal recorded a growth rate of 7 to 7.5% in the gross value added in agriculture compared to the national average of 3 to 4%. This agricultural growth we can say is unsustainable, as it has reached a peak due to the earlier technological upgradation. But, plan financial allocation to agriculture has been declining and meager funds are available for research and technological improvement.

The first round of green revolution was confined to a few crops (wheat, rice) and to few regions (Punjab, Haryana) etc. The yield improvement across crops was modest — which could compensate for the decrease in the rate of area expansion.

However, we must accept that the rate of agricultural growth over the decades, especially in the 1980s and 1990s remained ahead of population growth rate. Effective population control at this point of time would have helped India progress, provided agricultural growth also increases by adopting new technology and means. (For details on this issue see Rao & Hanumappa, 1999:A-133-138).

In fact, the average annual growth rate of value added per worker in agriculture during 1993-99 was less than that during 1987-94, whereas the corresponding growth rates in secondary and tertiary sectors in the recent period were higher than the respective rates in the earlier period.

It increased from Rs. 17.3 billion in 1990-91 to Rs. 18.3 billion in 1991-92 and to Rs. 19.8 billion in 1992-93. But it declined substantially to Rs. 15.7 billion in 1993-94 and then again rose to Rs. 18.8 billion in 1994-95 and declined to Rs. 17.2 billion in 1995-96.

The price of urea was reduced by 10%. The retention pricing system was abolished for phosphatic fertilizer producers, but was retained for urea plants. In the same year, a farmer's subsidy of Rs. 1000 per metric ton was introduced for DAP and MOP. But in June 1993, DAP subsidy of Rs. 1000 per ton continued only for domestically produced DAP, but was discontinued for imported DAP. In July 1996, this was raised to Rs. 3000 per tonne for domestically produced DAP, to Rs. 1500 per tonne for imported DAP, and MOP.

There is a proposal to further reduce subsidies being given to decontrolled fertilizers. The government benefits to the tune of Rs. 1250 crores due to increase in urea prices and Rs. 510 crores due of increases in urea prices and Rs. 510 crores because of the increase in the prices of DAP and MOP.

In fact, studies using the Head Count Ratio (HCR) — reflecting the minimal concept of poverty based on food deprivation — show that agricultural growth does reduce poverty in India.

There is a contradiction present in this issue, especially with regard to electricity power. The political parties (more specifically the Congress party) is promising free / heavily subsidized power to farmers, while the State Electricity Boards (SEBs) are financial stress and losses.

There should he some farmers in each area with surplus funds and such people can join together to establish a local agro-processing unit that shall cater to the needs of the surrounding villages. The government should encourage such enterprising farmers.

Please note that in the present context of individual farm cultivation, the costs for tilling the !and, sowing of seeds, fertilizer use, etc., shall be high. There is environmental damage as this pattern of agriculture is fairly intensive. However, co-operativization of agriculture in a particular village not only reduces the farmers' input costs, but also gives additional time to learn from other's experiences. Experience of Punjab in consolidation of land holdings has proved to be a success story, where between 1965-91 over 16,90,879 acres of land has been consolidated. This consolidation of holdings has opened new vistas of prosperity for farmers and agricultural development has been stimulated on a vast scale.

The government is withholding major stocks of food grains in Food Corporation of India (FCI) godowns, which are already flooded with stocks and ironically not many takers for the stocks. Government's restriction would be justified in the case of emergencies such as in 1998 when the prices of onions, salt, etc., shoot up to unproportional rates. Otherwise, it would only be adversely affecting the farmers.

It is important to note that the International Crop Research Institute for Semi-Arid Tropics (ICRISAT), Hyderabad has a very good water harvesting structures under the watershed development program and experiencing these watershed development in nearby villages also. Other agencies and the governments should learn from such experiences.

Note that although we recommend reducing land-ceiling laws, redistribution of land by lowering the legal ceiling on landholdings as politically not feasible and the implementation of top-down subsidies-oriented poverty alleviation programs is not likely to be effective, hence this suggestion.

These figures are arrived at by assuming that private consumption will grow at 5% per annum up to the year 2000 and at 6% per annum between 2000 and 2020. Rao and Radhakrishna (1997) project that the incidence of rural poverty will come down to 11% in 2010 and 3% in 2020 and in urban areas 4% will remain poor in 2010 and 1% in 2020.

3

Agriculture in India: Issues and Challenges

"Agriculture is the backbone of the Indian Economy"- said Mahatma Gandhi five decades ago. Even today, as we enter the new millennium, the situation is still the same, with almost the entire economy being sustained by agriculture, which is the mainstay of the villages. Not only the economy, but also every one of us looks up to agriculture for our sustenance too.

SIGNIFICANCE OF AGRICULTURE

Although agriculture contributes only 21% of India's GDP, its importance in the country's economic, social, and political fabric goes well beyond this indicator. The rural areas are still home to some 72 percent of the India's 1.1 billion people, a large number of whom are poor. Most of the rural poor depend on rain-fed agriculture and fragile forests for their livelihoods.

The sharp rise in foodgrain production during India's Green Revolution of the 1970s enabled the country to achieve self-sufficiency in foodgrains and stave off the threat of famine. Agricultural intensification in the 1970s to 1980s saw an increased demand for rural labor that raised rural wages and, together with declining food prices, reduced rural poverty.

Sustained, although much slower, agricultural growth in the 1990s reduced rural poverty to 26.3 percent by 1999/00. Since then, however, the slowdown in agricultural growth has become a major cause for concern. India's rice yields are one-third of China's and about half of those in Vietnam and

Indonesia. With the exception of sugarcane, potato and tea, the same is true for most other agricultural commodities.

The Government of India places high priority on reducing poverty by raising agricultural productivity. However, bold action from policymakers will be required to shift away from the existing subsidy-based regime that is no longer sustainable, to build a solid foundation for a highly productive, internationally competitive, and diversified agricultural sector.

ISSUES AND CHALLENGES

It is here the challenge arises considering the implementation of the technology at various levels in the Global community. The need of the hour is not application of the technology but the adoption of appropriate technology, which would suit the particular level of the global community. In India, the farming practices are too haphazard and non-scientific and hence need some forethought before implementing any new technology.

Applications of agricultural inputs at uniform rates across the field without due regard to in-field variations in soil fertility and crop conditions does not yield desirable results in terms of crop yield. The management of in-field variability in soil fertility and crop conditions for improving the crop production and minimizing the environmental impact is the crux of precision farming.

Thus, the information on spatial variability in soil fertility status and crop conditions is a pre-requisite for adoption of precision farming. Space technology including global positioning system (GPS) and GIS holds good promise in deriving information on soil attributes and crop yield, and allows monitoring seasonally- variable soil and crop characteristics, namely soil moisture, crop-phenology, growth, evapotranspiration, nutrient deficiency, crop disease, and weed and insect infestation, which, in turn, help in optimizing inputs and maximizing crop yield and income. Though widely adopted in developed countries, the adoption of precision farming in India is yet to take a firm ground primarily due to its unique pattern of land holdings, poor infrastructure, lack of farmers' inclination to take risk, socio-economic and demographic conditions.

FACTORS CONTRIBUTION TO DECLINE OF AGRICULTURE

Slow Down in Agricultural and Rural Non-Farm Growth: Both the poorest as well as the more prosperous 'Green Revolution' states of Punjab, Haryana, Andhra Pradesh and Uttar Pradesh have recently witnessed a slow-down in agricultural growth and it ultimately lead for farmer's suicide. Some of the factors hampering the revival of growth are:

- *Poor composition of public expenditures*: Public spending on agricultural subsidies is crowding out productivity-enhancing investments such as agricultural research and extension, as well as investments in rural infrastructure, and the health and education of the rural people. In 1999/2000, agricultural subsidies amounted to 3 percent of GDP and were over 7 times the public investments in the sector.
- *Over-regulation of domestic agricultural trade*: While economic and trade reforms in the 1990s helped to improve the incentive framework, over-regulation of domestic trade has increased costs, price risks and uncertainty, undermining the sector's competitiveness.
- *Government interventions in labor, land, and credit markets*: More rapid growth of the rural non-farm sector is constrained by government interventions in factor markets — labor, land, and credit — and in output markets, such as the small-scale reservation of enterprises.
- *Inadequate infrastructure and services in rural areas.* Infrastructure is also a significant factor in the process of development but country like our rural Bharat has not posses the infrastructure such as roads, electricity, fertilizer and pesticides availability which caused the vulnerable damage to the growth of agriculture.

Weak Framework for Sustainable Water Management and Irrigation:

* *Inequitable allocation of water*: Many states lack the incentives, policy, regulatory, and institutional framework for the efficient, sustainable, and equitable allocation of water.

Deteriorating irrigation infrastructure:

* Public spending in irrigation is spread over many uncompleted projects. In addition, existing infrastructure has rapidly deteriorated as operations and maintenance is given lower priority.

Inadequate Access to Land and Finance: Stringent land regulations discourage rural investments: While land distribution has become less skewed, land policy and regulations to increase security of tenure (including restrictions or bans on renting land or converting it to other uses) have had the unintended effect of reducing access by the landless and discouraging rural investments.

Computerization of land records has brought to light institutional weaknesses: State government initiatives to computerize land records have reduced transaction costs and increased transparency, but also brought to light institutional weaknesses.

Rural poor have little access to credit: While India has a wide network of rural finance institutions, many of the rural poor remain excluded, due to inefficiencies in the formal finance institutions, the weak regulatory framework, high transaction costs, and risks associated with lending to agriculture.

Weak Natural Resources Management: One quarter of India's population depends on forests for at least part of their livelihoods.

A purely conservation approach to forests is ineffective: Experience in India shows that a purely conservation approach to natural resources management does not work effectively and does little to reduce poverty. Weak resource rights for forest communities: The forest sector is also faced with weak resource rights and economic incentives for communities, an inefficient legal framework and participatory management, and poor access to markets.

Weak Delivery of Basic Services in Rural Areas: Low bureaucratic accountability and inefficient use of public funds: Despite large expenditures in rural development, a highly centralized bureaucracy with low accountability and inefficient use of public funds limit their impact on poverty. In 1992, India amended its Constitution to create three tiers of democratically elected rural local governments bringing governance down to the villages. However, the transfer of authority, funds, and functionaries to these local bodies is progressing slowly, in part due to political vested interests. The poor are not empowered to contribute to shaping public programs or to hold local governments accountable.

Measures Needed Areas

Enhancing Agricultural Productivity, Competitiveness, and Rural Growth

Enhancing productivity: Creating a more productive, internationally competitive and diversified agricultural sector would require a shift in public expenditures away from subsidies towards productivity enhancing investments. Second it will require removing the restrictions on domestic private trade to improve the investment climate and meet expanding market opportunities. Third, the agricultural research and extension systems need to be strengthened to improve access to productivity enhancing technologies. The diverse conditions across India suggest the importance of regionally differentiated strategies, with a strong focus on the lagging states.

Improving Water Resource and Irrigation/Drainage Management: Increase in multi-sectoral competition for water highlights the need to formulate water policies and unbundle water resources management from irrigation service delivery. Other key priorities include: (i) modernizing Irrigation and Drainage Departments to integrate the participation of farmers and other agencies in irrigation management; (ii) improving cost recovery; (iii) rationalizing public expenditures, with priority to completing schemes with the highest returns; and (iv) allocating sufficient resources for operations and maintenance for the sustainability of investments.

Strengthening rural non-farm sector growth: Rising incomes are fueling demand for higher-value fresh and processed agricultural products in domestic markets and globally, which open new opportunities for agricultural diversification to higher value products (e.g. horticulture, livestock), agro-processing and related services.

The government needs to shift its role from direct intervention and overregulation to creating the enabling environment for private sector participation and competition for agribusiness and more broadly, the rural non-farm sector growth.

Improving the rural investment climate includes removing trade controls, rationalizing labor regulations and the tax regime (i.e. adoption of the value added tax system), and improving access to credit and key infrastructure (e.g. roads, electricity, ports, markets).

Improving Access to Assets and Sustainable Natural Resource Use

Balancing poverty reduction and conservation priorities: Finding win-win combinations for conservation and poverty reduction will be critical to sustainable natural resource management. This will involve addressing legal, policy and institutional constraints to devolving resource rights, and transferring responsibilities to local communities.

Improving access to land: States can build on the growing consensus to reform land policy, particularly land tenancy policy and land administration system. States that do not have tenancy restrictions can provide useful lessons in this regard. Over the longer term, a more holistic approach to land administration policies, regulations and institutions is necessary to ensure tenure security, reduce costs, and ensure fairness and sustainability of the system.

Improving access to rural finance: It would require improving the performance of regional rural banks and rural credit cooperatives by enhancing regulatory oversight, removing government control and ownership, and strengthening the legal framework for loan recovery and the use of land as collateral. It would also involve creating an enabling environment for the development of micro-finance institutions in rural areas.

Strengthening Institutions for the Poor and Promoting Rural Livelihood

Promoting Community-Based Rural Development: State Government efforts in scaling up livelihood and community-driven development approaches will be critical to build social capital in the poorest areas as well as to expand savings mobilization, promote productive investments, income generating opportunities and sustainable natural resource management.

Direct support to self-help groups, village committees, user's associations, savings and loans groups and others can provide the initial 'push' to move organizations to higher level and access to new economic opportunities.

Moreover, social mobilization and particularly the empowerment of women's groups, through increased capacity for collective action will provide communities with greater "voice" and bargaining power in dealing with the private sector, markets and financial services.

Strengthening Accountability for Service Delivery: As decentralization efforts are pursued and local governments are given more prominence in the basic service delivery, the establishment of accountability mechanisms becomes critical.

Local governments' capacity to identify local priorities through participatory budgeting and planning needs to be strengthened. This, in turn, would improve the rural investment climate, facilitating the involvement of the private sector, creating employment opportunities and linkages between farm and non-form sectors.

THE HISTORY OF GARDENING: A TIMELINE FROM ANCIENT TIMES TO THE TWENTIETH CENTURY

A Timeline

From Ancient Times to 1600

35,000 BCE (BCE = Before the Common Era, or Before the Roman Era)

B.C. = Before Caesar (Julius Caesar), or Before Christ (Jesus of Nazareth)

Before the Communications Era of Handwritten Books and Scholarly Libraries

(450 BCE - 450 CE)

BCE does *not* mean "Before the Christian Era."

Evidence from archeological sites (tools, corprolites) indicates that Homo Sapiens at the end of the Paleolithic period had knowledge of many plants dervied from food gathering techniques. Different kids of fruits, nuts, and roots were only gathered, not cultivated.

15,000 BCE

"The history of the origin of human civilizations and agriculture is, of course, much older than the documentation in the form of pyramids, inscriptions and bas-reliefs or tombs can tell us. A close acquaintance with cultivated plants and with the multitude of types and their differentiation into geographical groups as well as their frequently sharp physiological isolation from each other compel us to refer the very origin of cultivated plants to such remote epochs, where periods of 5-10,000 years such as concern archeologists represent but a brief moment." ,- N.I. Vavilow, *Origin and Geography of Cultivated Plants*, 1987.

Dogs were used in hunting in Iraq.

8500 BCE

In Mesopotamia, humans raised domesticated goats, sheep, and cereal grains. Neolithic cultures involved farming.

"Scientists have carried out carbon-14 testing of animal and plant remains and have dated finds of domesticated sheep at 9000 BC in northern Iraq; cattle in the 6th millennium BC in northeastern Iran; goats at 8000 BC in central Iran; pigs at 8000 BC in Thailand and 7000 BC in Thessaly; onagers, or asses, at 7000 BC in Jarmo, Iraq; and horses at 4350 BC in Ukraine." Bio-Tech's History of Agriculture.

8000 BCE

"Certain cereals and pulses (legumes) were domesticated in very ancient times. In about 8000 BC in the Fertile Crescent of the Near and Middle East (present-day Syria, Iran, Iraq, Turkey, Jordan, Israel), wheats, barley, lentil, pea, bitter vetch, chick-pea, and possibly faba bean, were brought into cultivation by the Neolithic people. These crops spread from the point of origin. Archaeological evidence indicates that the wheats, and some of the legumes, had reached Greece by 6000 BC and evidence of their presence within that millennium has been found in the Danube Basin, the Nile valley, and the Indian subcontinent (Pakistan). Dispersal continued throughout Europe, the crops reaching Britain and Scandinavia in 4,000-2,000 BC." - *The New Oxford Book of Food Plants* [Vaughan 1997]

"The first evidence for plant domestication is approximately10,000 years old, but the first society in which people were primarily dependent on domesticated crops and livestock does not appear until about 6,000 years ago."

Khabur Basin farming in Syria.

"The only factor that can account for the irreversible and nearly uniform emergence of agriculture throughout the world is the grown of populations beyond the size that hunting and gathering would support." Mark Cohen's Thesis. [Heiser 1990]

7000 BCE

People in Central America cultivate corn and other crops.

People at Tepe Ali Khosh in Iran cultivated 'Emmer' and 'Einkorn' wheat. [Baker 1978]

Gardening myths

Chatal Huyuk is the largest Neolithic site in the Near East.

Rice cultivation in Yangtze Valley of China.

Apples cultivated in southwestern Asia.

6000 BCE

Evidence of cultivation of wheat (but not breadwheat), barley (naked, not hulled), and lentils were found in the Neolithic Greek cultures of Thessaly, Crete, and the Cyclades.

Evidence of rice cultivation at Ho-mu-tu in South China; and, at Ban-po-ts'un in North China.

Oranges cultivated in India and Tigris River Valley.

Cultivation of maize in Peru.

5000 BCE

Wild pod corn is cultivated in the Tehuacan valley in Mexico. [Baker 1978]

Millet grown along the Yellow River (Huang Ho) in China.

The Native people of North America inhabit river flood plains and cultivate crops.

Irrigation begins in the Middle East. [Heiser 1990]

Cotton grown in Mexico.

Domestication of some wild plants by people living in the Mississippi River drainage basin

Wine making in Iran.

4800 BCE

Archeological evidence from Tehuacan in south central Mexico shows that maize, squash, chili peppers, avocados, and amaranth were cultivated. [Heiser 1990]

Domestication of citrus species in various parts of the world. Sweet Oranges by Stephen Hui.

4500 BCE

Evidence of managed woodlands in Britain.

4000 BCE

"As in the case of the cereals, the legumes are amongst the oldest crops cultivated by the human race. Between the cereals and legumes there is a parallel domestication: wheat, barley, pea, lentil, broad bean, and chick pea in West Asia and Europe; maize and common bean in Central America; ground nut in South America; pearl millet, sorghum, cowpea, and bambara groundnut in Africa; rice and soya bean in China." - - *The New Oxford Book of Food Plants*, xviii, 1997, by J. G. Vaughan and C. A. Geissler.

Farming in Mesopotamia by Sumerians. Hittites.

Indus Valley agriculture is very extensive: wheat, peas, sesame seed, barley, dates, mangoes.

3900 BCE

Rice grown in Southeast Asia, Korat area of Thailand.

Ancient World Web Index, Ancient Scripts - Web Resources

3700 BCE

Uruk period of Sumerian agriculture.

3500

Egyptian agriculture using extensive irrigation techniques.

Egyptian garden art

Cotton growing and cotton textiles quite advanced in India, and reamained so until the 13th century.

Ancient Egyptian Horticulture and Agriculture

3000 BCE

Written manuals for the use of herbs in medicine existed in Mesopotamia (now Iraq) and in China. Herbal remedies were widely used by the ancient people.

Potatoes are cultivated in the Andes mountains of Peru.

Lost Crops of the Incas

"Of the two thousand or so species in the bulging genus *Solanum*, about 170 are tuber-bearers. Of the tuber-bearers, only eight are routinely cultivated and eaten by people, and most of these have stuck pretty close to home in the Andes of Peru. Only one has reached international stardom: *S. tuberosum*, commonly known as the potato. The potato probably originated in Peru, where indications are that it was domesticated over six thousand years ago by high-altitude-dwelling ancestors of the Incas."

Blue Corn and Square Tomatoes: Unusual Facts about Common Garden Vegetables. By Rebecca Rupp. Garden Way Publishing 1987.

Egyptians in the Nile Valley manufacturing and wearing cotton clothes.

Egyptian tomb paintings show walled gardens with fish ponds and fruit trees.

Carved water basin from Tello in Mesopotamia. [Hirst 1999]

Olives cultivated in Crete and Syria.

Farming in India.

2700 BCE

Rhubarb cultivated in China for medicinal purposes.

Egyptians used over 500 plants, wild and cultivated, for medicinal purposes. Egyptian wine making.

Chinese Emperor Shen Nung's plant classification lists.

Hemp cultivation in China.

2500 BCE

Rice was an important food in Mohenjo-Daro near the Arabian Sea, and in the Yangtze Basin in China.

Cotton was cultivated and its fibers spun and woven in Peru and the Indus Valley of Asia. [Baker 1978]

Figs, grape vines, pomegranates, and dates in cultivation in Egypt and Asia. The first garden art was probably decorated grape arbors [Gothein 1928]. Olive trees cultivated in Crete.

Farming in England.

2000 BCE

Native Americans are growing many varieties of corn, beans, squash, sunflowers, as well as using many wild plants as foods.

Egyptians making paper from the papyrus plant. Watermelon cultivated in Africa, tea and bananas in India, apples in the Indus Valley.

1750 BCE

The Hammurabic Code. Includes sections on maintaining irrigation canals and ditches, and property laws regarding gardens.. Sumerian "Farmer's Almanac."

1495 BCE

Queen Hatshepsut of Egypt imports trees from conquered territory in North Africa.

Farming in Ancient Egypt

One of the oldest surviving garden plans is for the garden of a court official in Thebes.

1300 BCE

Ramses II has apples cultivated along the Nile.

1167 BCE

Ramses III, Egyptian King, (1198-1167) benefactor to many grand temple gardens and public buildings.

1275 BCE

The Torah establishes rules for kosher food.

1000 BCE

Irrigation begins in in Mexico. [Heiser 1990]

Sacred Places: Trees and the Sacred.

Tiglath Pileser I, King in Mesopotamia, enthusiastic gardener

800 BCE

Peanuts cultivated in Peru.

Farming in Africa.

700 BCE

Works and Days by Hesiod.

540 BCE

Hanging Gardens of Babylon. Built by slaves and peasants directed by King Nebuchadnezzar II.

Sugar cane grown along the Indus River.

485 BCE

King Darius the Great (521-485) and his paradise garden in Persia.

440 BCE

Herodotus of Halicarnassos (484-426) writes on history, customs and life in the ancient world.

377 BCE

Hippocrates (circa 460-377) Greek physician. Wrote 87 treatises. Many herbal remedies.

350 BCE

Gardens at the Academy, Athens, Greece

Natural history references in the Jewish sacred scriptures- *Pentateuch*.

Trees: Living Links to the Classical Past. By John M. McMahon.

322 BCE

Aristotle (384-322) Greek philosopher and scientist. Wrote 26 treatises on natural science.

On Plants, Parts of Animals, On the Soul, Generation, Physics, On the Heavens.

Theophrastus inherited Aristotle's botanic garden in Athens, and many of Aristotle's treatises.

Books on plants and gardening written by Theophrastus. One of his books: De Causis Plantarum.

Theophrastus is considered by some to be the "Father of Botany."

Exchange of information, seeds and plants between Greece and Persia.

301 BCE

History of Plants and Theoretical Botany by Theophrastus.

Trees: Living Links to the Classical Past. By John M. McMahon.

Summary of Greek biology.

Rice growing in Japan.

271 BCE

Epicurus (341-271) used a large garden for gatherings and walks. The Philosophy Garden

Upon the Gardens of Epicurus; or, Of Gardening.By Sir William Temple, 1685.

207 BCE

The opulent and extensive gardens and palace of the first Chinese emperor Ch'in Shih Huang-ti were burned by peasants and Confucian rebels.

200 BCE

King Dutthagamini in India has a large artwork of the Sacred Fig Tree (Buddha's tree) made of precious materials and placed in the Great Gold Dust Dagoba park and gardens.

Gardens at Pompeii, Italy [Helphand 1977]

Greco-Roman eating, drinking, farming, farming and starving exhibit.

Almonds cultivated in Greece.

Cultivation and trade of coconuts between East Africa and India.

The Hopewellian farming culture in Ohio, North America, is thriving.

149 BCE

Cato (234-149) wrote on the simple country life.

De Agriculture, by Cato the Elder, emphasizes planting olives and grapes.

100 BCE

Grain harvesting at Karanis, Egypt

The Shang Lin (Great Grove) immense imperial garden of the Chinese emperor Wu-ti.

Shanlin Yuan ("yuan" is chinese for "garden") occupied over 1000 km^2 and contained more

87 BCE

The royal park and gardens of the Chinese Emperor Wu Ti (140-87) in West China, Chang-an.

The Roman's staple grain was spelt.

40 BCE

De Re Rustica. Varro (116-27). Roman agriculture. Varro was a prolific author, and he noted that there were over 40 known treatises available on the subject in 40 BCE.

29 BCE

Georgics. Virgil. Roman rural life.

Celtic Druids and Sacred Trees

50A.D. = "*Anno Domini*" in Latin or "*the year of the Lord*" in English.

Or...

CE= Common Era. The word "common" simply means that this is the most frequently used calendar system: the Gregorian Calendar. It does not mean "Christian Era."

*De Materia Medica.*Dioscorides the Greek. Herbal medicine.

60 CE

De Re Rustica, On Agriculture and Trees... Columella.

79 CE

*Natural History (Naturalis Historica).*Pliny the Elder (23-79). Roman naturalist.

90 CE

De Aquae Ductibus. Frontinus. Waterworks in the garden and farm.

105 CE

Tuscan villa at the base of the Apennies

Tai-yu's garden in China. Fiction by John P. Rastello.

113 CE

Pliny the Younger (61-113) Letters about villa gardens.

138 CE

Emperor Hadrian's Villa near Tivoli.

Karanis, a farming town in Roman Egypt.

250 CE

The administrators of the Roman Empire (circa 100 BCE-500 AD) actively exchanged information on agriculture, horticulture, animal husbandry, hydraulics, and botany.

Seeds and plants were widely shared.

Chinese making paper from rags, bark, hemp and other fibrous materials. [Baker 1978]

400 CE

The Palace Garden at Sigiriya in Sri Lanka

460 CE

Sidonius writes about his Roman villa in Lac d'Aydat in Auvergne, France.

Flora of Southeast Asia (Nan-fang ts'ao-mu chuang) by Hui-lin Li.

Chinese "scholar gardens."

Eggplants cultivated in China and India.

550 CE

Domestication of coffee takes place in Arabia until 800. [Baker 1978]

Coffee drinking popular in Arabia.

In the year 2000, coffee imports and exports are second only to oil on the world trade market.

560 CE

Ono No Imoko, Japanese Buddhist priest and scholar, living by a lake "ikebono", developed an elemental Ikebana flower arrangement style.

Mayan agriculture research

Various styles of fountains and water gardens reflect Persian, Indian and Roman influences.

618 CE

The Chinese emperor Yang-ti constructs the vast imperial garden called The Western Garden.

Suzho, China - "City of Gardens"; Pi Jiang Garden.

Arabs in Persia are impressed by gardening concept of *chahar bagh.*

670 CE

St. Fiacre - Patron Saint of Gardeners (620-670)

Sacred Trees of the Celts

735 CE

Venerable Bede, Saint Bede (673-735) English historian, scholar, and theologian.

De Natura Rerum - Medieval science. Many notes on monastic kitchen gardens.

750 CE

Use of the Green Man in art and lore becomes widespread in Christian Europe. Internet resources include: The Search for the Green Man and Who is the Green Man. Books on the subject include: *Green Man: The Archetype of Our Oneness with the Earth* by William Anderson, 1990. *The Green Man* by Kathleen Basford, 1978. *The Jack in The Green* by Roy Judge, 1979.

Arabs capture Chinese papermakers at Samarkand and adopt the process for papermaking.

760 CE

Hindu and Arabic mathematicians and thinkers are using a decimal arithmetic. Farmers and gardeners frequently keep detailed logs of their work, and decimal arithmetic is widely used to track important details, e.g., costs of plants and materials, percentage of plants in a batch of cuttings that took, quarts of berries picked, current supermarket price of fresh green beans, etc..

800 CE

The city of Baghdad is a center of Arab Islamic culture. Extensive scientific work on agriculture and botany for many centuries before and after.

Soft soap making widespread.

Chinese garden concepts and the Foolishman's Garden.

812 CE

Charlemagne (742-812) King of Franks, Emperor of Western Europe, patron of arts, sciences, and literature. Experimented with plants in a private garden and coordinated planting efforts on estates.

Plants in Medieval gardens.

850 CE

*Hortulus- Liber de Cultura Hortorum (Book Concerning the Cultivation of Gardens).*Walafridus Strabo (809-849).

Viking Age Foodstuffs

900 CE

Cordova, Moorish Spain, center for botanical studies and libraries and learning. Information.

Byzantine and Medieval Studies Links

Tofu commonly eaten in China.

Japanese gardening aesthetics and history.

1044 CE

The Great Hunger of 1044: The Progress of a Medieval Famine

Sacred Trees in Celtic Traditions

1050 CE

Tale of Genji. Japanese court novel describes aristocratic gardens.

A Chinese scholar's garden.

Medieval technology timeline.

1080 CE

*The Book of Agriculture.*Ibn Bassal, Arab botanist, plant collector, and horticulturist.

Farming in Dartmoor, England.

1085 CE

The great Arab libraries in Toledo, Spain, provide Europeans access to sophisticated Islamic and Greek writings in science and agriculture. The success of Arab agriculture in Andalusia, Spain, is renowned.

Horizontal looms are used in Europe.

1094 CE

Sakuteiki. Tachibana no Toshitsuna. Japan treatise on garden design.

1120 CE

The Chinese emperor Hui-tsung has the famous Ken Yeh Garden "The Impregnable Peak" constructed.

Manor system in Europe. A manor was roughly 900 to 2,000 acres of arable land.

Hard soaps in wide use.

1130 CE

Percussion drilling of wells in France.

1150 CE

The use of windmills for grinding grains.

1180 CE

Al-Awwam writing on Andalusian agriculture and garden design. Moorish Spain.

Ibn Baitar writing on medicinal plants: *Collection of Simple Drugs and Food.*

1191 CE

Tea from China becomes popular in Japan.

1227 CE

Vatican botanical garden founded. A medicinal or physic garden which still exists today, although in a different location.

St. Frances of Assisi (1182-1126). A holy man now known for his love of animals and nature, and his kindness.

1250 CE

The Japanese Buddhist priest Eisai (1141-1215) utilized a tea ritual as praticed in Chinese Buddhist temples.

Medieval European views about the spontaneous generation of organisms.

The wheelbarrow makes its way into European books.

1260 CE

De vegetabilibus. Saint Albert the Great. (1193-1280).

1280 CE

Marco Polo visits the palace garden of the Mogol ruler, Kubilai, in China.

Then he reports on visiting the famous Hsi Hu (West Lake) imperial gardens in the largest and probably most advanced city in the world at the time - (Kinsay) Hangchou, China.

He brought some new pasta making techniques back to Europe.

1305 CE

Opus Ruralium Commodorum, by Petrus de Crescentiis of Bologna.

Medieval agriculture.

1339 CE

Koki-dera (Moss Garden) of Muso Soseki, Japan.

Herding dogs used on European manors.

1350 CE

The great formal gardens of the Moorish Arabs (e.g., Generallife in the Alhambra, Granada, Spain) set standards.

Decameron. Giovanni Boccaccio. County gardens provide a retreat for those fleeing the plague.

1357 CE

The Black Death in Europe. A plague that reduced the population of Europe by 60%.

Alcazar gardens in Seville, Spain.

1390 CE

Forme of Cury: A Roll of Ancient English Cookery.

1400 CE

The Feate of Gardening.

1450 CE

*Illustrations for Designing Mountain, Water, and Hillside Field Landscapes.*Zoen, Japanese landscape architect.

Emperor Yoshimasa of Japan made flower arrangement part of universal education.

Johann Gutenberg began printing with moveable type in Mainz, Germany. By 1500, the world of ideas would never be the same in Western Europe because of this single invention.

Medieval gingerbread candy.

1460 CE

The Gart der Gesundheit. Printed in Mainz. Herbal medicine.

Hortus Sanitatis. Printed in Mainz. Herbal medicine.

Fifteenth Century Life in Europe - Roses

1470 CE

*De re aedificatoria.*Leone Battista Alberti (1404-1472). Renaissance scholar.

The Topkapi Palace in Turkish Constantinople has renowned fruit trees, gardens and landscaping.

Ottoman Turks introduce coffee to Constantinople. "The world's first coffee shop, Kiva Han, opens in 1475."

Pietro Creszenzi of Bologna compiled ancient works into *The Opús Ruralium Commodorum.*

Encyclopedia by Bartholomeus Anglicus. Apples and apple cider.

1480 CE

Giovanni Medici's villa garden.

Frankfurt-am-Main, Germany, may be the birthplace of the hot dog.

1490 CE

Temple garden of Royanjii, Japan.

1492 CE

Voyage of Christopher Columbus from Spain to the edge of the Americas.

The beginning of plant exchanges between Europe and the Americas.

Christopher Columbus: A Culinary History

1497 CE

Portuguese control the spice trade in the Indian Ocean.

1510 CE

Sunflowers from South America introduced in Spain.

1513 CE

Daisen-in garden in Koyoto, Japan. Designed by So-ami. This is a famous dry garden (*Kare-Sanuui*).

Hampton Court Gardens, England. By Linda Johnson.

1516 CE

First use of the term 'herbal' per the Oxford English Dictionary.

1528 CE

Hernando Cortes introduces vanilla beans, fava beans, cocoa, sweet potatoes, and haricot to Spain.

1529 CE

*Historia General de Nueva Espana.*Bernardino de Sahugun. Aztec gardening arts reported.

1530 CE

Gardens of Babur (1483-1530), Mughal Emperor, in Persia and India.

Persian botanical art, particularly miniatures, is renowned.

Plants exchanged between Europe and the Americas.

1533 CE

Oldest university chair of botany in Europe, founded in Padua by the Venetian Republic.

Spainards started caco tree plantations in Venezuela and Trinidad.

1535 CE

Nature Mysticism of Heinrich Cornelius Agrippa of Nettesheim (1487-1535) and Theophrastus Paracelsus (1493-1541).

1543 CE

Europe's first bontanic garden, established in Pisa by botany professor Luca Ghini.

Potatoes and tomatoes from South America, via Spain, cultivated in Europe.

1545 CE

The Ikenobo School "formulated the principles of rikka arrangements by naming the seven principal branches used in that type of arrangement."

1550 CE

Villa Medici in Rome.

Europe's first museum of natural history in Bologna.

The first printed almanacs in English become available.

1555 CE

Georgius Agricola [George Buaer} (1494-1555) German geologist, metallurgist, and paleontologist.

Carolus Clusius, Dutch botanist, cultivating tulip bulbs imported from Constantinople.

Fuch's Botanical. Leonhard Fuchs (1501-1566).

1557 CE

*Five Hundred Points of Good Husbandry: A Book of Huswifery.*Thomas Tusser.

Gonzalo Fernández de Oviedo y Valdes (1478-1557) Described flora of New World.

Bontanical book treasures in the Vatican Library.

1561 CE

De Historia Plantarum. Valerius Cordus.

1568 CE

William Turner (1510-1568), "Father of English Botany."

1569 CE

Nicolas Monardes writing about the botany of the New World based on Spanish accounts.

The Profitable Arte of Gardening.Thomas Hill. Herbal medicine.

Camembert cheese production in France.

1570 CE

Villa d'Este, Little Rome, constructed at Tivoli, Italy. Elaborate water garden.

Spanish explorers bring potatoes back to Europe.

Francisco Hernandez, private physician to Philip II of Spain, explores the New World and reports on over 1,000 plants considered of medicinal value. This research was not published until 1651 as *Rerum Medicarum Novae Hispaniae*.

The Enchanted Gardens of the Renaissance Facts about three Renaissance gardens near Rome: Villa D'Este - Tivoli, Villa Lante - Bagnaia, Bomarzo's Sacred Groves.

1576 CE

Conrad Heresbach (1496-1576) The Whole Art and Trade of Husbandry, Contained in Foure Bookes.

1577 CE

Gardener's Labyrinth.Thomas Hill (Didymus Mountain).

Sultan Murad III had the area of Anatolian Maras send him 100,000 hyacinth bulbs.

1580 CE

Villa Lante, Renaissance garden, Bagnaia, Italy.

1583 CE

Great Pharmacopoeia.Li Shih-Chen (Li Shi-Zhen). Chinese botanist. Botanical medicine.

De Plantis Libri. Andrea Cesalpino. A very important book in the history of botany. Plants grouped by physical characteristics (morphology) rather than by medicinal properties.

1586 CE

Sir Frances Drake brings sassafras and potatoes from America to England. [Rupp 1990]

Gardens in the Netherlands.

Vicino Orsini's garden at Bomarzo, Italy.

The Good Huswifes Handmaid for Cookerie in her Kitchen. Make a apple orange tart.

1591 CE

Sen no Soyeki or Rikyu (1522-1591). Japanese tea master, poet, and garden lover.

Tea. A great selection of teas and teaware; and some good information about tea.

1593 CE

First French botanic garden in Montpellier. Influenced by Moorish Spain.

1595 CE

Frances Bacon prepares lists of common garden plants.

Floriculture and plant collecting are very popular in England and the Low Countries.

1597 CE

The Herbal of Generall Historie of Plants. John Gerard. 1360 pages.

The Dutch take over from the Portuguese in contolling the spice trade in the Indian Ocean.

1600 CE

European forests are becoming depleted, and shortages of wood effect various industries. In later years, coal, petroleum, hydroelectric and finally nuclear power sources are increasingly utilized. [Ponting 1991]

Mannerism in gardening.

4

Impact of Globalization on Indian Agriculture

Agriculture, since its invention and inception, has been the prime and foremost activity of every culture and civilization throughout the history of mankind. Agricultural practices secured the food supply. It also resulted in establishment of farming settlements that subsequently grew in villages, towns, cities, states, countries and empires.

The history of agriculture is broadly divided in four phases, differing widely in date according to the region. The four phases are

1. *Prehistoric Agriculture*- Believed to have started near 10,000 BC or earlier, it is considered as first prototype of agricultural modules. The main agricultural sites were in southeast and southwest Asia, in what are now India, Iran, Iraq, Israel and the adjoining areas. Nile River in Egypt, Danube in Europe and Indus-Ganga in India were the main center of the agricultural farmlands.

   This was the age of agricultural innovations as many new tools and practices were invented due to compulsion of planting, sustaining and harvesting the crops. Use of wooden and earthen tools as well as some domesticated animals are major highlights of this period.

2. *Historical Agriculture*: This age roughly covers the period from 2500 BC to 500 BC. It marks the introduction of metals in agricultural tool making. Agricultural sites meanwhile grew proficient and covered much of

southeast Asia, middle-east, Europe, North America and eastern Asia.

Mechanicals improvements in agricultural implements and development of storage and irrigation method marks the highlight of this age. For the first time, trade in agriculture also started, becoming the precursor to modern agribusiness industry.

3. *Feudal Agriculture*: The feudal age of agriculture ranges from 500 BC to 1700 AD, reaching its height near 1000 AD and then getting static thereafter.

   This period is chiefly characterized by expansion of agriculture and irrigation and consolidation of agricultural methods. Use of fertilizers, in form of natural manures to boost production was a major breakthrough of this period.

   The usual practice of this age was of intensive agricultural throughout Europe, middle east and southeastern Asia.

4. *Scientific Agriculture*: By 17th century AD world was seeing a rapid rise in its population. This was also the age of imperial expansions in Asia, Africa, Americas and oceanic countries.

   These led to exploration of new methods and practices in agriculture. It also resulted in increased attention on commercial form of agriculture that was based solely on profit than subsistence purposes. Agriculture trade assumed global proportions and many new economies thrived on specialized production of one or other agricultural products.

Industrial revolution also led to many new inventions for agricultural sector, particularly in irrigation, sowing, and harvesting techniques. But it was the modern inventions in biotechnology and genetics that have completely changed the form of agriculture production. New variants of seeds and plants are manufactured in laboratories and genetically modified to suit particular consumption and commercial needs. This has enabled the world to meet the food demands of for a population that is crossing 6.5 billion mark.

The liberalisation of India's economy was adopted by India in 1991. Facing a severe economic crisis, India approached the IMF for a loan, and the IMF granted what is called a 'structural adjustment' loan, which is a loan with certain conditions attached which relate to a structural change in the economy. The government ushered in a new era of economic reforms based on these conditions. These reforms (broadly called Liberalisation by the Indian media) can be broadly classified into three areas: Liberalisation, privatization and globalization. Essentially, the reforms sought to gradually phase out government control of the market (liberalisation), privatize public sector organizations (privatization), and reduce export subsidies and import barriers to enable free trade (globalization).

There was a considerable amount of debate in India at the time of the introduction of the reforms, it being a dramatic departure from the protectionist, socialist nature of the Indian economy up until then. However, reforms in the agricultural sector in particular came under severe criticism in the late 1990s, when 221 farmers in the south Indian state of Andhra Pradesh committed suicide. (The damage done, 2005) The trend was noticed in several other states, and the figure today, according to a leading journalist and activist, P. Sainath1, stands at 100,000 across the country. (Sainath, 2006) Coupled with this was a sharp drop in agricultural growth from 4.69% in 1991 to 2.06% in 1997.

This paper seeks to look into these and other similar negative trends in Indian agriculture today, and in analyzing the causes, will look at the extent to which liberalisation reforms have contributed to its current condition. It will look at supporting data from three Indian states which have been badly affected by the crisis: Andhra Pradesh, Maharashtra and Kerala. Andhra Pradesh's (AP's) experience is particularly critical in this debate because it was headed by Chief Minister Chandrababu Naidu, who pursued liberalization with enthusiasm. Hence liberalization in AP has been faster than other states, and the extent of its impact has been wider and deeper. (Sainath, 2005)

Indian Agriculture Today: A Snapshot

Agriculture employs 60% of the Indian population today, yet it contributes only 20.6% to the GDP. (Isaac, 2005) Agricultural production fell by 12.6% in 2003, one of the sharpest drops in independent India's history. Agricultural growth slowed from 4.69% in 1991 to 2.6% in 1997-1998 and to 1.1% in 2002-2003. (Agricultural Statistics at a Glance, 2006) This slowdown in agriculture is in contrast to the 6% growth rate of the Indian economy for almost the whole of the past decade. Farmer suicides were 12% of the total suicides in the country in 2000, the highest ever in independent India's history. (Unofficial estimates put them as high as 100,000 across the country, while government estimates are much lower at 25,000.

This is largely because only those who hold the title of land in their names are considered farmers, and this ignores women farmers who rarely hold land titles, and other family members who run the farms.) (Sainath, P) Agricultural wages even today are $1.5 - $2.0 a day, some of the lowest in the world. (Issac, 2005) Institutional credit (or regulated credit) accounts for only 20% of credit taken among small and marginal farmers in rural areas, with the remaining being provided by private moneylenders who charge interest rates as high as 24% a month. (Sainath, 2005) An $NSSO_2$ survey in 2005 found that 66% of all farm households own less than one hectare of land. It also found that 48.6% of all farmer households are in debt. The same year, a report by the Commission of Farmer's welfare in Andhra Pradesh concluded that agriculture in the state was in 'an advanced stage of crisis', the most extreme manifestation of which was the rise in suicides among farmers. Given the performance of agriculture and figures of farmer suicides across the country, this can be said to apply to Indian agriculture as a whole.

THE CRISIS FACING INDIAN AGRICULTURE

The biggest problem Indian agriculture faces today and the number one cause of farmer suicides is debt. Forcing farmers into a debt trap are soaring input costs, the plummeting price of produce and a lack of proper credit facilities, which makes farmers turn to private moneylenders who charge exorbitant rates of interest. In order to repay these debts, farmers borrow again and get caught in a debt trap. I will examine each one these causes which led to the current crisis in Andhra Pradesh, Kerala and Maharashtra, and analyse the role that liberalisation policies have played.

As was mentioned earlier, AP's experience is particularly relevant in this analysis because of its leadership. Let me explain in detail. Chandrababu Naidu, Chief Minister of Andhra Pradesh from 1995-2004, was an IT savvy neo-liberal, and believed that the way to lead Andhra Pradesh into the future was through technology and an IT revolution. His zeal led to the first ever state level (as opposed to national level) agreement with the World Bank, which entailed a loan of USD 830 million (AUD 1 billion) in exchange to a series of reforms in AP's industry and government. Naidu envisaged corporate style agriculture in AP, and implemented World Bank liberalisation policies with great enthusiasm and gusto. He drew severe criticism from opponents, saying he was using AP as a laboratory for extreme neo-liberal experiments. Hence, AP's experience with liberalization is critical.

The Debt Trap and the Role of Liberalisation

The Debt Trap: High Input Costs

Seeds: The biggest input for farmers is seeds. Before liberalisation, farmers across the country had access to seeds from state government institutions. For example, AP's APSSDC3 produced its own seeds, was responsible for their quality and price, and had a statutory duty to ensure seeds were supplied to all regions in the state, no matter how remote. The seed market was well regulated, and this ensured quality in privately sold seeds too. (The damage done, 2005) With liberalization, India's seed market was opened up to global agribusinesses like Monsanto, Cargill and Syn Genta. Also, following the deregulation guidelines of the IMF, 14 of the 24 units of the APSSDC's seed processing units were closed down in 2003, with similar closures in other states.

This hit farmers doubly hard: in an unregulated market, seed prices shot up, and fake seeds made an appearance in a big way. Seed cost per acre in 1991 was Rs. 70 (AUD 2) but in 2005, after the dismantling of APSSDC and other similar organizations, the price jumped to Rs. 1000 (AUD 28), a hike of 1428%, with the cost of genetically modified pest resistant seeds like Monsanto's BT Cotton costing Rs. 3200 or more per acre, (AUD 91) a hike of 3555%. (Sainath, 2005) BT Cotton is cotton seed that is genetically modified to resist pests, the success of which is disputed: farmers in Andhra Pradesh and Maharashtra now claim that yields are far lower than promised by Monsanto, and there are fears that pests are developing resistance to the seeds. (Menon & Jayaraman, 2002) Expecting high yields, farmers invest heavily in such seeds. Also BT Cotton and other new seeds guarantee a much lower germination rate of 65% as opposed to a 90% rate of state certified seeds.

Hence 35% of the farmer's investment in seeds is a waste. (Sainath, 2004) Output is not commensurate with the heavy investment in the seeds, and farmers are pushed into debt. The abundant availability of spurious seeds is another problem which leads to crop failures. Either tempted by their lower price, or unable to discern the difference, farmers invest heavily in these seeds, and again, low output pushes them into debt. Earlier, farmers could save a part of the harvest and use the seeds for the next cultivation, but some genetically modified seeds, known as Terminator, prevent harvested seeds from germinating, hence forcing the farmers to invest in them every season.

Fertilizer and Pesticide: One measure of the liberalisation policy which had an immediate adverse effect on farmers was the devaluation of the Indian Rupee in 1991 by 25% (an explicit condition of the IMF loan). Indian crops became very cheap and attractive in the global market, and led to an export drive. Farmers were encouraged to shift from growing a mixture of traditional crops to export oriented 'cash crops' like chilli, cotton and tobacco. (The damage done, 2005) These need far more inputs of pesticide, fertilizer and water than traditional crops. Liberalisation policies reduced pesticide subsidy (another explicit condition of the IMF agreement) by two thirds by 2000. Farmers in Maharashtra who spent Rs. 90 an acre (AUD 2.5) now spend between Rs. 1000 and 3000 (AUD 28.5 – 85) representing a hike of 1000% to 3333%.

Fertilizer prices have increased 300% (Sainath, 2005) Electricity tariffs have also been increased: in Andhra Pradesh tariff was increased 5 times between 1998 and 2003. (Seeds of ruin, 2005) Pre-liberalisation, subsidised electricity was a success, allowing farmers to keep costs of production low. These costs increased dramatically when farmers turned to cultivation of cash crops, needing more water, hence more water pumps and higher consumption of electricity. Andhra Pradesh being traditionally drought prone worsened the situation.

This caused huge, unsustainable losses for the Andhra Pradesh State Electricity Board, which increased tariffs. (This was initiated by Chandrababu Naidu in partnership with Britain's DFID4 and the World Bank.) Also, the fact that only 39% of India's cultivable land is irrigated makes cultivation of cash crops largely unviable, but export oriented liberalisation policies and seed companies looking for profits continue to push farmers in that direction. (Isaac, 2005)

The Debt Trap: Low Price of Output

With a view to open India's markets, the liberalization reforms also withdrew tariffs and duties on imports, which protect and encourage domestic industry. By 2001, India completely removed restrictions on imports of almost 1,500 items including food. (The damage done, 2005) As a result, cheap imports flooded the market, pushing prices of crops like cotton and pepper down. Import tariffs on cotton now stand between 0 – 10%, encouraging imports into the country.

This excess supply of cotton in the market led cotton prices to crash more than 60% since 1995. As a result, most of the farmer suicides in Maharashtra were concentrated in the cotton belt till 2003 (after which paddy farmers followed the suicide trend). (Hardikar, 2006). Similarly, Kerala, which is world renowned for pepper, has suffered as a result of 0% duty on imports of pepper from SAARC5 countries. Pepper, which sold at Rs. 27,000 a quintal (AUD 771) in 1998, crashed to Rs. 5000 (AUD 142) in 2004, a decline of 81%. As a result, Indian exports of pepper fell 31% in 2003 from the previous year. (Sainath, 2005) Combined with this, drought and crop failure has hit the pepper farmers of Kerala hard, and have forced them into a debt trap. Close to 50% of suicides among Kerala's farmers have been in pepper producing districts. (Mohankumar & Sharma, 2006)

The Debt Trap: Lack of Credit Facilities and Dependence on Private Money Lenders

In 1969, major Indian banks were nationalized, and priority was given to agrarian credit which was hitherto severely neglected. However, with liberalisation, efficiency being of utmost importance, such lending was deemed as being low-profit and inefficient, and credit extended to farmers was reduced dramatically, falling to 10.3% in 2001 against a recommended target of 18%. (Seeds of ruin, 2005) A lack of rural infrastructure deters private banks from setting up rural branches, with the responsibility falling on the government, which has reduced rural spending as a result of its liberalisation policies.

Rural development expenditure, which averaged 14.5% of GDP during 1985 – 1990 was reduced to 8% by 1998, and further to 6% since then.

This at a time when agriculture was going through a crisis proved disastrous for farmers, who turned to private money lenders who charge exorbitant rates of interest, sometimes up to 24% a month. (Seeds of Suicide, 2005) With input costs and output prices being what they are, coupled with crop failures and drought, they are pushed into debt which is impossible to repay. 12 out of India's 28 states have 50% and higher indebtedness among farm households.

Andhra Pradesh has the highest percentage of indebted farm households — 82%. 64.4% of Kerala's farm households and 54.8% of Maharashtra's farm households are indebted (NSSO, 2003) Indebtedness has been identified as the single major cause of suicides in both Andhra Pradesh, Kerala and Maharashtra. (Analysis of Farmer Suicides in Kerala, 2006, Who's suicide is it anyway, 2005, Saving small farmers, 2005)

Liberalisation and How it Failed

Branco Milanovic, a World Bank economist describes how he believes liberalisation helps developing countries achieve growth: 'when a country lowers trade barriers, reduces government intervention in the market in order to allow market forces to operate freely, increases competition and attracts foreign investment, it will increase productivity and reduce inefficiency, which will lead to economic growth, and in a few generations, if not less, the poor will become rich, illiteracy will disappear, and poor countries will catch up with the rich.' (The damage done, 2005) This argument is an economic rationalist one, which views government intervention with profound suspicion, and has equally profound faith in unfettered market forces. (Whitwell quoting Robert Manne, 1998) What Mr. Milanovic neglects to mention, though, is that rich countries which now preach liberalisation protected their 'infant industries' at the time they began to industrialize, till they were strong enough to compete globally.

The US government, for example, had a protectionist trade policy in the late nineteenth century to help US companies become competitive in the world.

Besides, apart from wool, the US, Germany, Britain and France were all almost self-sufficient in the raw materials that they needed for industrialization, and took off from that platform, a luxury that India and other developing countries do not have. (Issac, 2005) As German economist Friedrich List says, the adoption of these values (of liberalisation) assumes that all countries are at the same starting place, which (as we have seen above) is clearly not the case. (Issac, 2005) In fact, it is this very reason that has brought about the crisis that Indian agriculture is facing today.

Most farmers in India were already in a position of minimum security, with no education system, credit facilities, access to alternative employment, or efficient technology. Their only support was government subsidy and regulation. Liberalisation policies came in and dismantled their only support structure. It halted the sharp reduction in rural poverty from 55% in the 1970s to 34% in the 1980s. Not only has the incidence of poverty in rural areas not gone lower than 34% in the 1990s, it has gone to higher levels of 42% in individual years. (Ghosh, 2000)

The second most popular argument of the economic rationalists in favour of liberalisation is that competition will weed out the inefficient, and in the growth that ensues, employment will be provided in other areas of the economy, thus lifting the poor out of poverty. This argument however assumes that the poor will be able to take advantage of the opportunities presented to them. As Robert Issac says in 'The Globalization Gap', "Globalization encourages the well positioned to use tools of economics and politics to exploit market opportunities, boost technical productivity, and maximize short-term material interests." This is compounded in India, where the gap between one who is 'well positioned' and one who is not can be extreme. With a lack of investment, chances of generation of rural employment are slim.

Unemployment and underemployment are chronic problems in India, with the rate of unemployment being close to 10% in 2004. (Sainath, 2005) Primary education in rural areas is mismanaged and bad quality, and there is no system which helps agricultural workers find alternate employment, or develop alternate skills. (Chossudovskly, 1997) In the face of such obstacles, it is nearly impossible to expect agricultural workers to shift to alternate fields. Coming back to AP, the IT Revolution spearheaded by Chandrababu Naidu attracted companies like Google, Amazon, Microsoft and Dell and created thousands of jobs. However, given the skills and education of most farmers, it is obvious that none of this translated into job opportunities for them.

The final argument that supporters of globalization have is the much touted 10% reduction in poverty (60 million decline in poor) in India in the year 2000. However, this figure was challenged by experts. Poverty is defined according to how many people consume less than the nutritional minimum prescribed. (2400 calories for rural areas, and 2100 for urban areas) Major changes in survey design in 1999-2000 not only made the resultant estimates incomparable to previous years' estimates, but an over-estimation of consumption (meaning people were getting enough food, hence were not considered poor) meant a sharp reduction in poverty figures. After experts challenged it, the Planning Commission of India accepted that the figure was inaccurate, and could not be compared to previous years' estimates, hence the 10% drop in poverty is incorrect. With adjusted figures, experts have determined that the decrease in poverty was a mere 2.3%, and that the number of poor increased by nine million in 2002 as compared to 1999.

Liberalization and 'Growth'

Many economists now concede that the relationship between liberalisation and growth are 'uncertain at best'. According to the Center for Economic and Policy research, which studied impact of liberalisation reforms on the developing world, key economic and social indicators such as increases in life expectancy, infant and child mortality, education and literacy levels slowed down in the 20 years between 1980 and 2000 when liberalisation policies were implemented, compared to the 20 years leading to 1980. (The damage done, 2005) This defeats the economic rationalist argument of free trade eliminating poverty, since the 20 years leading up to 1980 witnessed high protectionist policies and trade barriers. Following the suicides in 2000, the World Bank and Britain's DFID abandoned power reforms in Andhra Pradesh four years before schedule. It admitted that it had 'substantially underestimated' the 'complexity of the process' and that there must be 'increased consultation with the farmers to get their acceptance' of any further reform. (The damage done, 2005).

The Andhra Pradesh government sponsored report by the Commission of Farmer's Welfare squarely laid the blame for its agrarian crisis on the state and central government's policies: "While the causes of this crisis are complex and manifold, they are they are dominantly related to public policy. The economic strategy of the past decade at both central government and state government levels has systematically reduced the protection afforded to farmers and exposed them to market volatility and private profiteering without adequate regulation; has reduced critical forms of public expenditure; has destroyed important public institutions, and has not adequately generated other non-agricultural economic activities." A report on suicides in Kerala similarly held the liberalization policies of the government responsible. (Mohankumar & Sharma, 2006)

CONCLUSION

It is clear that the liberalisation policies adopted by the government of India played a dominant role in the agrarian crisis that is now being played out. However, this is not to say that privatisation, liberalisation and globalization are per say bad, or inherently inimical to an economy. It is the 'one size fits all' brand of liberalisation adopted by the IMF and the World Bank which forces countries to privatize, liberalise and globalize without exception which has failed. Without taking into account the state of an economy, and in this case, the state and nature of the agricultural sector in India, the IMF and the World Bank, with the cooperation of the Indian government, embarked on mismatched reforms, which have caused misery and despair among millions of Indian farmers, driving large numbers of them to suicide. It is also essential to break the link between aid and liberalisation, which caused India in the first place to accept the conditions of the IMF.

Remember that India was on the brink of a financial crisis in 1991 when it applied for the IMF loan and accepted its conditions—perhaps the course of economic reform in India would have taken a very different course if there was no urgent need to borrow from the IMF. The start to this process may have already occurred: recognizing the failure of its liberalisation policies, (and perhaps also the failure of DFID with AP's power reforms) the Blair government of Britain announced in 2004 that it will no longer make liberalisation and privatization conditions of aid. In another blow to the neo-liberal lobby, Chandrababu Naidu suffered the worst ever defeat in the 2004 state elections in his party's history, with rural AP clearly rejecting his brand of World Bank sponsored liberalisation. The battle, however, has not yet been won. It is essential for the rest of the G8 to follow Britain's example in order to influence World Bank and IMF policy towards India to ensure blind liberalization is not pursued, and so that countries like India can adopt tailor-made reforms to suit their economy.

MAKING INDIA AN ECONOMIC SUPER POWER

For a nation like India, politically subjugated for centuries by alien conquest, and socially unsettled by invasions, the idea of becoming a super power is no ordinary thought. Understandably, even after winning freedom, India never aspired seriously to become a super power, politically or economically. Such an idea, if it existed, was dormant in the Indian consciousness. But it did acquire seriousness after the Pokhran blasts in May, 1998.

But even now, the national opinion is not uniform on how to become a super power. "Not atom bombs, only development can make India a super power," say some. "We need both to make us a super power", say others. Nevertheless, the idea that we should aim at being a super power is no more a dormant one, even if it is not a dominant one. But the desire has come a little late. There is background to it — the background of India's post-Independence history.

The freedom movement had promised much more than a mere change of rulers. But what happened was, as aptly described by English historians, a mere "transfer of power" from the aliens to the locals. Nevertheless, what did India intend to be after it became free?

When India attained freedom, it thought of emerging as a global leader, without becoming a global power. Its claim to leadership rested on the age-old Indian values of universal tolerance, peace and happiness. But the post-War world afflicted by cold war had no reverence for such high values. So India was swiftly marginalised in a world which respected only power.

But, within India, the Indian leadership did the other way round — it persuaded the people not to pursue their age-old values, but, accept the Anglo-Saxon ideas and institutions in the main. It folded back the philosophic lead shown by Gandhi, Aurobindo and Tilak. Their definition of the Indian identity was substituted by the western ideas of secularism and socialism. Since then, for over four decades, the Left-Socialist parties and intellectuals mounted a vicious attack on the Indian past, and virtually delinked the Indian polity, economy, history and education from its past and turned to Anglo-Saxon values.

This is precisely what the Indian freedom movement had struggled against. Because of this drift, to the Indian intellectual, India's past became a burden and ceased to be matter of pride. Also, the secular socialist leadership systematically fragmented the Indian society into majority and minority, rich and poor, forwards and backwards, and denied India the deeper awareness of its intrinsic unity brought about by Hindu values and civilisation. This resulted in setting one Indian against another, leading to massive self-deprecation. This eroded our self-confidence as a nation. The idea of a powerful India could not be internalised in a situation where every Indian was running down the other.

The socialist regime turned two generations of potential entrepreneurs into job-seekers. Permit quota and licence, not efficiency or merit, became the route to business success. The competitive strength of India was systematically weakened, and the traditional skills of the trading and business communities in India were dissipated. The devastation caused by the socialist regime in the post-Independence period was more pervasive than the devastation of the Indian economy by the British rule.

The pre-British India, described as caste-ridden, feudalistic and unmodern, was economically ahead of the rest of the world,including Britain and the USA. The Indian economy had a share of 19 per cent of global production in 1830, and 18 per cent of global trade, when the share of Britain was 8 per cent in production and 9 per cent in trade, and that of USA, 2 per cent in production and 1 per cent in trade. India had hundreds of thousands of village schools and had a functioning literacy rate of over 30 per cent. In contrast, when the British left, India's share of world production and trade declined to less than 1 per cent and its literacy was down to 17 per cent. And yet, in 1947, India had large Sterling reserves, no foreign debt, and Indians still had an effective presence in such trade centres as Singapore, Hong Kong, Penang, Rangoon and Colombo. But, by the time the Socialist regime came to a close, India had become politically and economically weak and disoriented, lacking in self-confidence. Its influence in South Asia too had waned.

But, with the collapse of the Soviet Union, the Indian establishment swung to the other extreme — from socialism to market capitalism. It was again in tune with its lack of self-confidence — if it is not communism, it has to be capitalism. And we began to witness a massive effort to globalise — which means westernise — the Indian economy. The persons and parties who grossly maladministered the national economy suddenly turned 'reformers' and they paraded themselves as the new messiahs as if someone else was responsible for the wrongdoings in the past which needed to be reformed. Thus, Congressmen, that is those who needed to be reformed, styled themselves as 'the reformers'. Just as socialism was an ideology earlier, globalisation too became an ideology. Now anyone who dissents against globalisation is being labelled as backward-looking and anti-modern. Thus, just as we adopted socialism without a debate as to its viability in India and discarded it, we have adopted globalisation without any debate. As a result, we have set out blindly to experiment on India the experience of the West. And just as we had committed one set of mistakes before 1991, we have now begun committing another set of mistakes from 1991.

The great question remains: Will what we have been doing since 1991, make India an economic super power? If not, by what method will India emerge as an economic super power? Before we comprehend what India should do to become a global power, we must understand properly what is happening in the world today.

There is a gradual shift in the global view on global trade and currency regimes of today. It questions not just the global assumptions, but also the national assumptions about the global regime. The experience of Mexico first, the EC currency crisis next, and the South Asian crisis now, tend to question the very foundation, stability and viability of the present global economic regime under installation. The assumptions about what India should do is also changing fast. For instance, not even the diehard lobbyist for globalisation today talks about the full convertibility of Indian currency for which P. Chidambaram had, only a year ago, unveiled a three-year master plan. Dr Jagdish Bhagwati, an uncompromising advocate of currency convertibility, has now come to the view that currency convertibility should be "avoided like a plague". This single shift is enough to demonstrate how unstable are global economic prescriptions and institutions.

It is not that what happened in South Asia now was not anticipated. Akio Morita, the former Chairman of Sony Corporation, wrote a letter to the G7 leaders in 1992 questioning the rationale of globalisation with a currency regime operated by mindless speculation. Akio Morita said that what manufacturing, efficiency and technology achieved might be destroyed by the free-for-all money market over which no one had any control. The world was forced into the present free-for-all currency regime in 1971. It was to prevent this, and to prevent money turning as a destabiliser that the world opted for a fixed rate of exchange in 1949. The very constitution of the IMF, the World Bank and GATT after the World War II was to avoid a free-for-all currency regime. But it was reborn in the year 1971 after the gold-dollar convertibility was discontinued. But, with global economics coming under money-spinners, the new regime has been increasingly questioned, more seriously now.

What is the lesson for India in the changing situation? The WTO, which is the symbol of free trade, cannot function without a free currency market, but, a free currency market destabilises free trade. This is a Catch-22 situation for which there is no answer within the global trade regime today. A mere comparison between the trade volumes in the world currency market and the volumes of trade in goods and services is adequate to demonstrate how paper money dominates trade. The daily global trade in currency is over $ 1.2 trillion; in contrast, the annual global trade in goods and services is $ 4.5 trillions. It means that for every dollar of actual trade, the trade and speculation in currency is $ 95!

If anything that will ultimately destroy the global trade regime, it will be the global currency regime. That means the present global trade regime does not have a guaranteed success, though it is not guaranteed to fail. India should not reorganise its national economy to match a global regime which has no guarantee for success. It means that India shall have to frame a national agenda.

So, the road to economic success in future will be not unlimited globalisation, but, controlled, guided, calibrated globalisation which is the vehicle for acquiring global economic strength. Even in the past, it has been so. Japan became an economic power by controlled globalisation even as China is doing the same thing now. It means that India has to devise its own strategy, that is what the idea of Swadeshi is all about. Swadeshi means Indian agenda to which the global agenda can only be a supplement, and not the other way round. This is what China did in the 1980s, it set out on a national agenda to double its food production from 200 million tonnes to 400 million tonnes between 1983 and 1993. It turned to its agriculture to power its economic growth beyond 10 per cent during the 1980s and turned to other areas to sustain the growth rate afterwards.

In contrast, the Indian agriculture virtually remained stagnant throughout the period from 1990 to 1998. This is how we malprioritised our economic strategy even after we set out to desocialise our economy in 1991. What then should we do now?

The answer is obvious. If we have to break into a high trajectory growth rate on a sustained basis, just as China did, we have to start with agriculture. Higher agricultural growth leads to rural prosperity which, in turn, leads to more buying power, and that triggers high industrial growth. The Indian agriculture has greater advantages than the Chinese agriculture had. We have one-and-a-half-times as much plain land as China has and a whole year's sunshine. Agriculture involves the greatest number in our country and unless a nation activates a majority of its people, it cannot grow. Thus, agriculture is our biggest asset. We should have invested more and more capital in agriculture. But, in contrast, the capital formation in Indian agriculture is declining from year to year down from 17 per cent in 1980 to 9 per cent now. So agriculture in India has been increasingly neglected. Agriculture is the starting point for India to emerge as an economic super power.

The second trigger for high growth is the energy sector. It is linked to transport. We did not plan for multimodel transport, nor did we integrate energy and transport in planning. We have allowed the cheaper Railways to decline in favour of costly road transport. The Railways share in freight traffic has come down from 88 per cent in 1950 to 37 per cent now; in passenger traffic from 68 per cent to 13 per cent. The reduction in the Railways share has meant a corresponding rise in the share of energy — intensive road transport. Why did the Railways decline? It is because we have saddled our Railways, which is one-and-a-half-times in length of the US Railways and twice as that of China, with coal transport and coal accounts for 45 per cent of railway freight today. It completely clogs the railway system and prevents its modernisation. We have not yet tapped the vast potential of coastal transport and inland water transport which can considerably ease the pressure on the energy-intensive road transport and the clogged railways.

Water transport involves less than 1/3 of the energy as land transport. We have not also tapped atomic energy. The expert view is that by using fast-breeder reactors with international safety standards, India can produce 3,00,000 MW of atomic power with indigenous resources. Our current power generation from all sources is about 89,000 MW. Atomic energy has higher initial investment, but, no running cost. But in thermal power, running cost is as high as 50 per cent of the production cost.

In other countries, atomic energy constitutes a high percentage of total energy — it is 76 per cent in France, 40 per cent in Sweden, 29 per cent in Japan and 23 per cent in USA, but less than 2 per cent in India. Also we can turn alcohol into diesel like Brazil does, in large quantities. An integrated energy-cum-transport planning can trigger a high growth.

The third area for large-scale activation of India is the non-corporate sector which accounts for 40 per cent of our GDP, over 30 per cent of exports and more than 75 per cent of employment outside agriculture. This sector consists of the traditional craftsmen and artisans in the main. During 1997-98, this sector had contributed over $12 billion by way of remittance from skilled Indian workers employed abroad. From 1980, the total remittance by this sector has exceeded $ 75 billion. This is largely self-employed and labour intensive. This has withstood the socialist assault and, later, the free market globalisation.Yet this sector has had no assistance from the system by way of organised finance or employment or social insurance schemes. This has enormous potentiality for high growth.

As an adjunct to this is the services sector — both modern and traditional services. This is an area of high growth potentiality for India because of the advantage of familiarity with the English language, particularly in the modern services sector like law, accounting, consultancy and advertising. And yet we are mindlessly allowing the MNCs to enter this area where we have our dominance.

Yet another area in which India is expected to become a major player is textiles, particularly after the quantitative restrictions are removed by the West from 2005. Textiles account for 30 per cent of India's exports and the largest industrial employment. But this crucial area is underdeveloped in terms of technology and processing. Powerlooms, which account for 73 per cent of the textile produced in India, are not funded by organised financing. Organised finance, technical upgradation and freedom from tax inspectors will transform the textile sector into a high growth area.

Computer software and IT technology are an important area. The Indians in the Silicon Valley in the USA have proved our proficiency in IT products. The global players are sourcing India in a big way because of the highly skilled manpower available here. Some of the best minds in India are taking to this area of expertise. Since information is key to the world ahead, the Indian mastery over this area is likely to make India a dominant IT power in the next decade. There are estimates that by 2002, India will export software products worth $ 10 billion.

This takes us to the role of foreign direct investments in India emerging as a global power. By its very nature, FDI has a marginal, but, crucial role to play. But it must be made to flow into areas where the local capital is shy or incapable of funding a mega venture, particularly for global use. It can usefully supplement our large national savings which, in dollar terms, amount to over 90 billion. Instead of going for FDI in areas which hurt the local economy and employment, the government may allow 100 per cent FDI in, say, four large international airports or two large seaports. Such facilities require income to be generated from abroad for their very survival. Indian capital cannot build them and the ordinary Indian need not pay for them. Such airports alone can fetch an FDI of 20-30 billion dollars in the next five years. The thrust of the FDI regime must be to channel FDI where local capital is inadequate and from where the FDIs cannot leave India at will.

Indians must realise, as have the westerners and the Japanese, that global business is war, not just commerce. Wars have always followed commerce and commerce has followed wars. Therefore, trade, like war, is a game of deception; and free trade is a strategy, not an ideology. That is why nations build up non-tariff barriers. South Korea used to tax-raid the buyers of foreign cars. The Philippines mandates that cars may be produced by foreign companies, but, distribution can only be by a local company.

France has instituted professional appraisal of imported electronic goods so as to clear only a fraction of the imports.

Thus, every nation digs trenches to fight the advances of global trade to the prejudice of national interests. But in India, just as socialism did, free trade has emerged as an ideology. Indians must learn to look upon trade as war and free trade as a strategy and develop native non-tariff formulations.

Again, what India and Indians critically need is a measure of national pride and self-confidence. Without both, no nation can prosper. It is unfortunate that the English-educated Indians have repeatedly failed to generate personal or national self-confidence. Instead, they began to deprecate everything Indian. In the process, they deprecated each other, resulting in all-round self-deprecation. This trend has to be reversed. If we deprecate ourselves, as we have been habituated to, no one is going to respect India. The atomic test by India triggered our national pride in a big way, but we threw cold water on it. We have to build, and not waste, such achievements whether it is the Param Super Computer, or PSLV launch, the Tata Car or Kamini nuclear reactor working on thorium. A rapid build-up of national pride and self-confidence will yield a quantum jump in economic advancement. Trade is a matter of psychology.

When Japan followed the Chinese lead in technology, it prided itself as combining the 'Japanese spirit' and 'Chinese technology'; when it was forced to accept the western technology, it coined a new motto, 'Japanese spirit with western technology'. Thus, a measure of nationalism has to be built up to create a psychology of national pride. This is needed for economic advancement and those who delink national pride from economic growth will have to be educated on what the world history teaches every nation.

And, finally, what India needs, as the first step, to transform into a global economic power is national consensus. The confrontationist politics and political untouchability that we are witnessing for the past few decades has made the formation of national consensus difficult. But this is what the Japanese considered as the first step for national economic development. In fact, even the Japanese democracy is unlike the Anglo-Saxon model. It refused to accept the divisive idea of majority-minority party politics and continues to pursue, even today, consensus as the national principle.

But the Indian parliamentary democracy, which is a carbon copy of the western system, does not allow the emergence of national consensus in politics; even though the Indian society functions on consensus, the Indian polity does not. For transforming India into a global power expeditiously India has to review its polity and bring it in tune with its traditional principle of panchayat or the idea of consensus to which the Indian society is accustomed.

There is no need for a trade-off between democracy and development, as Lee Kwan Yew, the former President of Singapore, proposed. Probably, a presidential form of government will ensure the support that national economics needs from national politics. Yes, India has all the potentialities and also the compulsion to become a global economic power in the 21st century, and early enough in the coming millennium.

All that India expects of its leadership is that it should ensure that there are no round pegs in square holes. The leadership must inspire the nation to think and function as a nation. The essential quality of a leader is that he does not care how long will he be in power, but how decisively he rules. This is what Lal Bahadur Shastri did. He ruled for just 18 months and asked the people to starve on Mondays — a trans-religious auspicious day. He is remembered for both. India is expecting such a trait in its leadership to lay its claims to becoming an economic super power. Will the present leadership, which claims to be different from the previous ones, respond to the will of the nation to become a global economic power?

CIVIL SOCIETY & FARMERS' REPRESENTATION ON AGRICULTURE

"Organic agriculture is defined as a holistic food production management system, which promises and enhances agro-ecosystem health, including biodiversity, biological cycles and soil biological activity. It emphasizes the use of management practices in preference to the use of off-farm inputs, taking into account that regional conditions require locally adapted systems. This is accomplished by using, where possible, agronomic, biological and mechanical methods, as opposed to using synthetic materials to fulfill any specific function within the system." — FAO/WHO, Codex Alimentarius Commission

Agriculture in India is in crisis. The failure of the 'Green Revolution Technology' (GRT) is clearly evident in stagnant agricultural production, a mounting spiral of ecological problems, relentlessly rising input costs, and increasing farmer indebtedness and suicides. From the small, peasant farmer to the FAO, there is global consensus that the GRT path is unsustainable. It impoverishes both the farmer and the natural resource base of agriculture, and provides toxic food and water to consumers. The increasing billions of dollars spent every year to subsidize 3% of the population engaged in farming in the USA, is stark proof of its economic bankruptcy. Large corporates who provide agro inputs, are the real beneficiaries of chemical based agriculture.

India's 'National Commission on Farmers' recently reported: "40% of Indian farmers would like to leave farming if it is possible to do so". This summarizes the enormity of the present agricultural crisis and the challenge facing the nation: how to safeguard agricultural incomes or provide alternative livelihood support to a quarter billion people who are potential future economic and ecological refugees – uprooted by mounting farm production costs and a rapidly degrading natural resource base.

While some corporate interests are now lobbying for Genetically Modified (GM) species as the solution, a growing body of scientific testimony and evidence – from both India and round the world – warns against the many serious and irreversible dangers these pose. Such hazards include the uncontrolled and unwanted spread of certain genes and genetic traits; emergence of resistant and more virulent secondary pests; potential health hazards; and other unpredictable problems. Evidence of the economic counter-productivity of GM in the long run, is already emerging. In China, for instance, which has been using Bt cotton for over 7 years, there has been a severe rise of secondary pests when the bollworm is controlled, resulting in the same levels of pesticide spraying as before the use of Bt seeds. In India, there is an alarming record of an increase in crop failure, farmer suicides and deaths of grazing cattle after the use of Bt cotton in rainfed areas of Andhra Pradesh and Vidharbha, Maharashtra. The high cost, high risk and ecologically damaging GM technology is not the solution for our agrarian crisis, and will further damage Indian agriculture.

Both GRT and GM based agriculture require chemical intensive and fossil fuel intensive inputs, which directly contribute to greenhouse gases and climate change: an estimated 25% of the world's carbon dioxide emissions, 60% of methane gas emissions and 80% of nitrous oxide emissions. In fact nitrous oxide, produced by the use of nitrogenous fertilizers, is 200 times more potent than carbon dioxide as a greenhouse gas. Chemical based agriculture estimatedly uses ten times more energy as input than the calories it produces as food. Shifting to bio-diverse, organic systems can contribute significantly to mitigating global warming and climate change. National vulnerability to the rising costs of fossil fuels, inescapable with chemical farming, is also much reduced.

Small and marginal farmers, whose land holdings are below 2 ha, constitute almost 80% of all Indian farmers, and more than 90% of them are dependant on rain for their crops. With GRT, their costs of cultivation and risks of crop failure are so high that often the farmers cannot recover even the money spent. Between 1990-91 and 1995-96, chemical fertilizer costs increased by 113%, and pesticides by 90%, whereas the wholesale price of wheat went up only 58%. Minimum Support Prices for all crops, except sugar, were 38% to 50% lower than the actual cost of production. Per capita food grain production has fallen to levels lower than the 1939-44 famine. Despite the thousands of crores spent on fertilizer and other subsidies, farmers are increasingly in debt and despair. Apart from the necessary immediate reliefs in crisis areas, through write off of loans/ interest, and appropriate Minimum Support Prices for agricultural produce, there is urgent need for a planned and vigorous promotion of low-cost, low-risk, high nutrition, holistic and sustainable farming systems to reinvigorate Indian agriculture, and to stem the rising tide of farmers' indebtedness, distress and suicides.

It is the experience of increasingly large numbers of farmers that holistic farming systems, based on scientifically proven techniques, are a very successful remedy to the economic and ecological crises engendered by GRT Holistic farming systems rely on available natural resources and rebuild the ecological capital on which all agriculture is dependent.

They also greatly reduce or totally eliminate the dependence of farmers on the purchase of expensive external inputs such as seeds, fertilizers, pesticides and herbicides. Particularly for rain fed farming and horticulture, which cover at least 60 % of our cultivated land, holistic farming should be adopted on a large scale without fear of loss of production.

Depending on the extent of prior damage by chemical-intensive monocultures, there is some drop in productivity in the transition phase if adequate biological inputs are unavailable.

But within 2 to 3 years, the system is equally, or more, productive than other systems – and on an improving and sustainable growth path. This has been scientifically validated internationally by FAO reports, and by ICRISAT Patencheru (O P Rupela & others). Better nutrition (balanced and free of toxins), and distributive justice (helping the poorest) are additional important benefits of a holistic approach.

There is clear scientific empirical evidence that a holistic, organic approach to farming nurtures numerous effective microbes occurring in nature that make all the essential nutrients available to crops, and improve vital physical qualities of the soil, namely its porosity, aeration, moisture absorption, drainage, and resistance to erosion. Crop yields thus increase in a sustainable manner, input costs fall and farm incomes rise.

Mixed cropping – central to holistic farming – greatly increases the total yield of biomass, which in turn, absorbs more carbon dioxide and thus ameliorates global warming. It also helps maintain soil fertility through on-farm recycling, heightens security against climatic vagaries and crop failures, and improves the nutritional balance of local diets.

This will also help in reducing the large quantities of pulses and oilseeds imported every year.

With water and energy shortages fast mounting, it is imperative to cut wastages and improve their use efficiency. Holistic farming optimally utilizes available moisture, and vastly improves national water and energy efficiency.

Scientists estimate that for every 1% of organic matter content, the soil can hold 16,500 gallons of 'plant available water' per acre of soil of one foot depth (Source ATTRA) Groundwater recharge is also enhanced by increased biomass and mulch, correspondingly diminishing run-offs and floods.

11 POINT AGENDA FOR THE 11TH PLAN

Considering the above compelling reasons for adopting a holistic approach favoring sustainability, national health and distributive justice, we strongly recommend and petition the Government of India to adopt an 11 point programme on agriculture for the 11th Plan as follows:

(1) Holistic sustainable farming should be promoted vigorously to cover at least 20% of all farmland progressively every year, so that the major part of the cultivable area of India is converted to holistic, ecologically and economically sound and sustainably productive farming systems by the end of the 11th Plan. Government policies and programmes should be specifically prioritized, designed and targeted to help the small and marginal farmers, who constitute almost 80% of farmers, to achieve food and livelihood security. Even small holdings have demonstrated that they can efficiently meet food and livelihood needs with proper policy and support.

(2) The current excessive use of inorganic nitrogenous fertilizer that is incrementally destroying the vital biological quality of Indian soils, and contributing to global warming, should be discouraged by phasing out the subsidy on urea through a progressive 20-25% reduction each year, publicly announced in advance. Similarly, subsidies for all other agro-chemicals and heavy farm machinery should also be phased out within 5 years.

(3) All the money saved by reducing/withdrawing subsidies as aforesaid (Rs 22000 crore per annum for urea alone) should be used as incentive/support for farmers who adopt holistic organic farming This should include subsidy for local generation of bio-fertilisers, including the planting of trees for enhanced availability of biomass and nutrients. Each farmer adopting holistic, organic agriculture, helps build up the nation's agricultural capital of soil and water, but faces some risks and costs in the initial 2 to 3 years of transition period commonly needed for ecological and economic recovery. Soil-depleting chemical fertilizers have been subsidized all these years, and it is therefore warranted that a similar subsidy should be extended to those who produce their own fertilizers and enhance soil capacity for sustainable productivity. A cash incentive to holistic farmers of Rs. 4,000 per ha per year for 2 or 3 years is strongly recommended to encourage this vitally necessary shift.

(4) A country-wide campaign must be undertaken on a war footing to regenerate our natural resources of soil, water, biodiversity, tree/forest cover, etc. Natural forests, grasslands and wetlands must be conserved and regenerated as they shelter fauna, avifauna and insects vital for agriculture, besides creating new fertile soil and augmenting recharge of groundwater. Outstanding benefits can thereby be efficiently achieved in the shortest possible time with significant savings in expenditure.

(5) Decentralised water harvesting and its optimal and equitable use must be actively promoted and financed by Govt policies, programmes and fiscal incentives/ disincentives. Since at least 60% of Indian farming is rain-fed, the small and marginal farmer is at the mercy of an increasingly erratic monsoon. The first priority should be protective irrigation, through full subsidy for small farm ponds and/or ground water recharge by farmers This will reduce monsoon crop failure and facilitate introduction of a winter crop, thereby greatly reducing risks and increasing agricultural production and farmer incomes. This can be financed by decreasing the proposed massive and wasteful expenditures on large irrigation projects, having long gestation periods, high inefficiencies, fiscal irregularities and high social displacement, rehabilitation and ecological costs. Farmers need immediate provision of water on-farm, not through interlinking of rivers. Where irrigation projects exist, water delivery must be regulated through suitable local bodies like Water Users Societies, SHGs (Self Help Groups) and/or Rural Producer Companies controlled by locally resident farmers. A differential, incremental pricing mechanism should be mandatory for higher per capita /per acre levels of consumption of surface or ground water in excess of the minimum/basic allotment. (The water used to irrigate one acre of sugarcane is sufficient for at least 25 acres of jowar, bajra or maize.)

(6) The main objective of EGS, grants, subsidies, loans and Govt schemes should be asset creation for the small and marginal farmers. Delivery should be channeled through local SHGs, associations, or bodies of farmers. These, as well as Producer Companies that are wholly owned and controlled by locally resident farmers and which can engage the services of professionals to handle marketing, finance and other support services, must be actively promoted and supported. Such support/incentives to local farmers' bodies should be proportionate to the

number of farmer members represented by each such recipient body, with additional benefits to those consisting exclusively of small and marginal farmers primarily engaged in natural resource regenerating holistic agriculture.

(7) The financial capacity of small Indian farmers, who are at the mercy of the market, should be strengthened so that they can hold their surplus produce till the prices are remunerative, and be helped to take up activities to add value to their produce through basic processing, packing, and collective marketing. Some Cooperative Banks, which provide grain storage and loans to farmers against grain stocks, are examples to be up-scaled. Government should promote the formation of local bodies/ Producer Companies of farmers and help them meet their needs for working capital, storage structures, community banks for grain, seed, fodder and other biomass, basic food and biomass processing units, tree planting, rainwater harvesting, biodiversity regeneration, extension and education, subsidy on small machinery/farming tools/aids, oxen, bullock carts, etc. The Government has long provided generous incentives for the promotion of industry. Bodies of small and marginal farmers engaged in holistic farming should similarly receive tax waivers/concessions, low interest loans, and easier access to credit. There should be a farmer-friendly, single window delivery system for credit and Govt schemes through the rural banking network, which must be rapidly extended to support farmer groups. Organic Certification must be provided free. In fact, certification of chemically cultivated and GM crops, and cautionary labeling of those that are packaged, should be mandatory in the interests of consumers.

(8) Low cost and nutritious food must become available to all, and particularly to the poor. The Public Distribution System should offer a wide variety of locally produced nutritious crops and be managed by local SHGs or Producer Companies under civil regulation. This will

reduce the high fuel cost and carbon dioxide generation of long-distance transportation. There should be total liberalization for primary processing, free movement and regional marketing of farm produce within a 150-200 km radius from farm to retail outlet, and proactive government support for basic transportation and regional marketing infrastructure which shortens the supply chain between farmers and consumers.

(9) Good Agricultural Practices used by innovative holistic farmers should be widely, systematically and urgently disseminated by use of Govt institutions, media, personnel and funds. Since the knowledge base for holistic farming systems is to be found among experienced, innovative farmers, a partnership should be forged between them and existent educational, research and extension facilities to document, use and disseminate this knowledge. This can be accomplished faster by granting/leasing 10% of the land of Agricultural Universities and KVKs to innovative farmers with at least 3 years of demonstrated success in holistic farming for demonstration and/or extension of low cost holistic farming systems, including agriculture, horticulture, livestock and rainwater management, etc. Agricultural scientists and extension workers should help in scientifically documenting and widely disseminating this information.

(10) Genetically modified organisms/seeds must be banned in India – or, at the very least, strongly discouraged through the most stringent regulation and heavy, deterrent penalties for any spread of genetic pollution. They are likely to create new and more damaging cycles of ecological and economic risks. A handful of multinationals and Indian corporates are inducing farmers to adopt this high-cost, high-risk, poorly regulated technology by offering credit and promises of short-term benefits. Consumers across the world have been rejecting GM food and increasingly demanding poison free organic food. This is India's opportunity to be a

leading producer of organic food and avail of this fast growing domestic and global market. The promotion of GM and GRT destroys this capacity and must be immediately halted and then reversed.

(11) Policies which permit the dumping of heavily subsidized agricultural products in India by developed countries – facilitated by the WTO – or which promote cash crop monocultures for export, making Indian farmers vulnerable to the disastrous consequences of any fall in global prices, must be discontinued.

Promoting holistic farming systems - which can provide food security to the 650 million Indians dependent on agriculture and nutritious food to all consumers- can effectively counter this threat to the nation's food security and farmers' livelihood and survival. It should therefore be the keystone of our agricultural policy, practices and priorities.

CONCLUSION

Holistic farming involves knowledge, native wisdom, and labour, rather than external industrial inputs or large per capita finance - with their attendant high costs and risks of indebtedness.

It can create significantly higher employment opportunities in the rural sector, thereby halting and reversing migration from rural to urban areas. It benefits the small and marginal farmers, who are the " Aam Admi" of the farming community. Government funds and resources should therefore be made available to them in proportion to their percentage in the rural population to honour the promises made in the Common Minimum Programme of the UPA government. Expropriating their land, the only asset of small and marginal farmers, by the Government for SEZs, and providing huge fiscal incentives to their promoters, is a policy that would inevitably lead to further pauperization of the farming community and the growth of armed insurgence.

Empowering farmers to ensure their own food and livelihood security through holistic farming systems, and through dispersed small industry based on agricultural produce, is the real need of the hour and the demand of the farmers, who constitute 65% of India's population.

All over the world, the demand for toxin-free produce is rising at an exponential rate of growth. Consumers in India are also entitled to it. Many States, and the farmers themselves, would be keen to adopt holistic farming systems if necessary support, as stated above, is made available. The Planning Commission, the Government of India, and the leaders of all political parties will fail in their duty if appropriate actions are not taken urgently to halt the promotion of GRT and GM agriculture, and to introduce the low-cost, low-risk, holistic farming systems which can save the small farmers from desperation and suicide. We thus earnestly call for your support in implementing the above proposals in the interests of the farming community and the nation as a whole.

5

India-US Knowledge Initiative on Agriculture Work Plan

The Board met on 13th – 14th February, 2006 in New Delhi, India. After the opening remarks by the Co-Chairs of the two countries and presentation of the Indian and US proposals, it was decided to identify the items of convergence under the 4 priority areas of co-operation for developing a work plan indicating the short and longer term deliverables. In its deliberations, the Board had the benefit of the guidance of Professor M.S. Swaminathan, Honorary Adviser to the Board.

Four Working Groups held detailed deliberations and presented their recommendations for the Work Plan under the Knowledge Initiative.

The Work Plan agreed to by the Board is described below:

(i) Under Education, learning resources, curriculum development and training, the proposals on building, human and institutional capacity are included. Mechanisms include establishing a standing Indo-US Joint Working Group, sharing the US experience in curriculum development to benefit the ongoing curriculum revision exercise in India, identifying the number of slots for training and faculty exchange during 2006 with consideration of additional slots in the coming years, organizing workshops for review and planning in certain identified areas, etc. With regard to building institutional capacity, the focus is on extension and outreach activities, library resources, networking and leadership development through sharing of experiences to take the knowledge initiative forward. The modalities for implementation of the training and other activities were also finalized.

(ii) Under Food-processing and Marketing, the Work Plan targets training, capacity building and joint research, including quality assurance and food safety, reduction of post harvest losses, market information systems, value addition, strengthening grades and standards, facilitation of agro-business investment, advanced processing technologies, byproduct utilization and bio-fuels from bio-mass.

(iii) Under Biotechnology, a strategic alliance has been envisaged for training and research on development of transgenic crops with resistance to economically important viruses, tolerance to drought, heat and salinity and micro-nutrient utilization efficiency; molecular breeding and genomics in crops and animals, molecular approaches in plants and animal health protection have been agreed.

(iv) Under Water Management, research and training proposals on sustainable use of water resources, water quality management and remediation, use of modern tools in water management, integrated nutrient management and precision agriculture have been decided.

Funding: The proposed Work Plan will be supported by an allocation of Rs. 3500 million (about USD 80 million at current rate of exchange) by the Indian Government over a period of three years. The allocation from the US side will be determined.

The initiative is expected to offer a win-win situation for both the countries and will trigger benefits in perpetuity.

BACKGROUND

Pursuant to the understanding on cooperation in Agriculture reflected in the Joint Statement of US President George Bush and Indian Prime Minister Dr. Man Mohan Singh on July 18,

2005 during the visit of Dr. Man Mohan Singh to the USA, the Ministry of Agriculture, Government of India and U.S. Department of Agriculture (USDA) agreed to work together for a new India – US Knowledge Initiative on Agriculture Education, Research, Services and Commercial Linkages, identified the objectives of this Initiative and created the Board of the Knowledge Initiative on Agriculture (AKI) for identifying areas of common interest, modalities of action plan and funding mechanism, etc. In the first meeting of the joint India-US AKI Board held on 15th & 16th December, 2005 in Washington, D.C., the Board agreed on the following key areas:

(a) Human Resources and Institutional Capacity - Building (cross cutting areas):
- Education, learning resources, curriculum development and training
- R&D, commercialization
- IPR, biosafety, food safety, investment conditions, and policy and regulatory frameworks

(b) Agri-Processing and Marketing
- Food processing
- Use of byproducts and bio-fuels
- Post-harvest management/value addition
- Food marketing, cold-chain/product handling

(c) Emerging Technologies
- Biotechnology
- Nanotechnology
- Nutraceuticals
- Vaccines and diagnostics
- Precision farming
- Bioinformatics and information technology

(d) Natural Resources Management
- Water management

- Soil and management
- Climate change
- Air quality Waste management

The Board further agreed that among the above identified areas, it would initially concentrate on (1) Education, learning resources, curriculum development and training (cross cutting), (2) Food processing, use of byproducts and bio-fuels, (3) Biotechnology, and (4) Water Management.

As agreed above, the Secretariat on both sides of AKI Board developed detailed proposals in early 2006 in consultation with experts and stakeholders in their respective countries. The Proposals so prepared by both countries were exchanged and discussed at length during the 2nd Board meeting. As a result of above deliberations, the following Work Plan for the AKI has been developed under the agreed priority areas.

Education, Learning Resources, Curriculum Development And Training: Building Human and Institutional Capacity

Agricultural education needs to be globally competitive. The contribution of US collaboration in the past to the Human Resource Development activities is widely acknowledged in India.

The present program envisages collaboration and mutual benefit by strengthening human resource development and research capabilities in niche/cutting edge areas for developing technologies relevant to national, regional and global needs.

In the changed context, importance has to be given to attaining excellence in agricultural education, to enhance employability of the graduates, developing human resources to fulfill commitments towards the Millennium Development Goals, and improving quality of life through sustainable rural development including innovative agricultural extension, agri-business programmes and wider participation of women.

With the rapid change in the technology environment, there is demand for preparing graduates to harness science and technology for the pursuit of attaining and sustaining the 'Evergreen Revolution'. This calls for new courses as per changing market demand and new modes of delivery with effective use of electronic media. Efforts will be made to modernize course curricula and methods of delivery i.e., moving from teaching to learning. This would bring about substantial improvements in the tools and techniques of imparting education in that teacher and taught are more interactive and learning is for upgrading skills in real life situations.

A critical mass of faculty and scientists in subject domains and need based niche/strategic/emerging areas is required to be developed through post-doctoral programs, faculty exchange, trainings, workshops etc.

These select faculty and scientists, in turn, would serve as catalysts for further change and improvement. The proposed faculty and scientists exchange would reinforce the education and research by offering need-based courses employing innovative delivery mechanisms and initiating research with national, regional and global relevance.

The new information and communication technologies (ICTs) would enable reaching the un-reached anywhere, any time and at competitive cost. To make this most effective, exposure of faculty and scientists to the ICT experiences to develop need-based entrepreneurship oriented programs assumes significance.

There is need to develop learning resources. Faculty and scientists need to be trained to develop resources using multimedia, web based technologies and in the transmission and retrieval of digital resources. The digital resources can be made available across National Agricultural Research System. Access to digital learning resources would enhance quality of teaching, extension and research across the NARS institutions.

The Work Plan will have the main following objectives:

- To enhance quality and relevance of higher education through reorientation and refinement of course curricula, learning resources and delivery processes.

- To develop and enhance human capacity with due focus on gender balance in the emerging areas through training and faculty exchange
- To promote industry-academia interaction to enhance relevance of education and research on a changing time scale.

WORK PLAN

Curriculum Development

- Establish a standing India-US Joint Working Group, by October, 2006, for capacity building in curriculum development, design and delivery with emphasis on need – based, demand driven and market-oriented agriculture, and relevant tools such as in-service training and distance education
- The Indian Council of Agricultural Research (ICAR) is in the process of course curriculum revision. To benefit the ongoing exercise, the US shall share its recent experiences in curriculum development by April, 2006 with ICAR.

Training and Faculty Exchanges

Training (In-service and Group) and faculty exchanges: This activity will include in-service training and participation in collaborative teaching/research. The duration may be up to 6 months.

- The areas of training under the identified areas include
  - (i) under education, learning resources, curriculum development and training: instruction, design and development, distance learning, teaching technology,
  - (ii) under food processing, use of by-products and bio-fuels: extraction technology, extrusion technology, food quality and safety, fermentation and enzyme technology, byproduct utilization, and fuel alcohol from biomass and market institution building and agri-business investment

(iii) under biotechnology: transformation in plants, molecular diagnostics and vaccines, allele mining, gene therapy and gene silencing and

(iv) under water management: water quality management and remediation, water-poverty mapping, soil/plant water relationships, marine population dynamics

- Under the cutting-edge/niche areas, the key training areas include climate change, carbon sequestration and trading, nanotechnology, protected and precision farming, bio-prospecting, valuation of biodiversity, biosafety, rural development, livelihood systems, gender issues, agri-business management, international trade or any other area to be identified.
- For these trainings, 24 slots shall be considered in 2006 and additional slots to be considered in subsequent years.
- Further, under (i) biotechnology, (ii) water management, and (iii) food processing, additional training slots shall be considered for 2006 and the following years.

Public - Private Sector Partnerships

- *Professorial chairs*: The industry/private sector sponsored chairs to be created in India or USA for R&D on strategic/niche areas. This will help in close collaboration between public and private sectors, which in turn will lead to commercialization of technologies at a faster pace.
- *Industry-Academia Interface Workshops*: Each year, faculty (premier US and Indian business /management schools and agri-business institutions) and industries will be invited to a Workshop to devise strategies, by which synergies can be induced to explore emerging trends and needs for the agriculture sector in both countries and orient education, training and research contributing to economic growth. This workshop may be held in India or the USA, and may be cosponsored by

the private sector stakeholders. The workshop will inventory, upgrade and build on existing agri-business programmes to match students or professionals with practical internship experiences.

Building Institutional Capacity

(a) EXTENSION AND OUTREACH

- Jointly organize a workshop in India to discuss innovations, e-extension, global lessons learned and best practices relevant to the needs of India's agricultural sector by March/April, 2007.

(b) LIBRARY RESOURCES

- Establishment of a virtual partnership (National Agricultural Library, Beltsville and National Agricultural Library, New Delhi) towards strengthening the library resources and electronic library networks in India relevant to the agricultural knowledge system (subject to copyright regulations).

(c) STRENGTHENING ADMINISTRATION

- Leadership development – involving institutional leaders and administrators to create a Research and Education Managers' Partnership by early 2008 that will share experiences, and focus on leading, managing and taking the Knowledge Initiative forward.

Modalities for Implementation of the Hrd Programmes

- The salary, travel cost, and health insurance for trainees and faculty to be borne by the sending country.
- The receiving country is to provide per diem (lodging & boarding), training cost and bench fees.
- In the case of post-doctoral programmes (up to two years), the sending country will provide the salary and the travel and related costs; receiving country/institution will provide fellowship support.

- The above conditions may be varied with mutual agreement.
- For Workshops, the host country will meet the expenses related to organization of the event and the sending country will meet the travel and related costs of its participants.
- The right to publish and seek protection for any innovations/technologies would be in accordance with an agreement between the collaborating institutions and funding agencies.
- The number of slots offered with adequate consideration given to gender in allocating the slots will be mutually agreed upon subject to availability of funds.

Participating Institutions

Considering the needs in niche/emerging/cutting-edge areas, selected institutions from the NARS (viz. agricultural universities, CAU, Deemed Universities, ICAR Research Institutes, participating private sectors etc.) will participate.

Based on the past experience in collaborative efforts with Indo-US initiatives, established capabilities, lead in niche/emerging/cutting edge/strategic research, future potential, and the capability enhancement of different institutions of Indian National Agricultural Research System shall be undertaken in a phased manner.

The probable partners from India have been identified in the Indian proposal. The US side will also identify the US institutions for partnership in each activity.

Food Processing, Use of by-products and Bio-fuels

The AKI Board agreed that the development of agricultural marketing and processing industries is now a priority for India's increasingly need-based and demand driven, market oriented agricultural sector. The Board agreed to work together on a range of capacity building initiatives and research activities in a framework of public-private partnership, wherever needed, and promotion of linkages between technology and the market

- Post Harvest Management and Upgrading Cold Chain Practices and Operations: Training and exchanges on improvement of the quality and safety of perishable food products would include temperature-related food-safety and product-quality-degradation issues faced by distributors, wholesalers and retailers. Training would also address reducing microbial contamination and maintaining constant product temperatures. In addition, U.S. private-sector and university experts and potential investors in cold chain development would visit India to meet counterparts, survey existing infrastructure, and identify weaknesses in cold chain practices and constraints to investment. This 3-phase activity would commence in late 2006 and carry on through 2007.
- Strengthening Grades and Standards and Quality Control: This would consist of training, exchanges and workshops, with involvement of the private sector and public sectors, on commodity grades and standards, the operation of grading and marketing programs, and inspection systems to maintain quality of crop and animal products. This ongoing activity would continue through 2006-2007 and would be focused on specific crops and products.
- Supporting Development of Market Information Systems: This would include training and exchanges on institutional capacity building related to market data collection, processing, analysis and timely distribution of market information to farmers, agribusiness, and policymakers. This would build on the ongoing USAID/ USDA projects in this area and continue through 2006 –2007.
- Facilitating Agribusiness Investment in India: This would be comprised of one or more sector-specific "Trade and Investment Mission/s to India" in 2007-2008 for U.S. private agri-business companies to meet counterparts, survey agri-business and identify opportunities and constraints faced by investors. The mission(s) could be preceded by an agribusiness

investment-related conference in India in 2007 and technical exchange programs focused on relevant issues.

- Drawing on the U.S. Experience with Contract Farming: Exchanges and workshops in 2006-2007 would explore the development of legal mechanisms for contracts, alternative approaches to regulating contracts, and specifying contracts to meet the needs of buyers / sellers (farmers).
- Establishing Food Safety and Animal and Plant Health Regulations To Facilitate Trade: The United States and India would initiate cooperative activities to better understand each country's Sanitary and Phyto-sanitary (SPS) systems, Technical Barriers to Trade (TBT) and food safety regulations. These activities would focus on consistency with relevant international norms and bringing further efficiency / transparency into thc regulatory processes. The cooperative activities would include consultations with all stakeholders including the private sector.

JOINT RESEARCH PROGRAMMES

- Technology for rapid detection and control of bio-toxins, chemical contaminants and heavy metals in agricultural produce and by-products: Food quality and safety are essential for both domestic and export markets. Development/ acquisition of rapid test equipment and protocols for ensuring food quality at various points in the value chain would be developed through trainings and joint applied research programmes.
- Advanced Extrusion Processing Technology: To undertake training of scientists and applied research projects in the area of extrusion processing for development of fortified and nutritious food products.
- Advance technologies for extraction of oils, oleoresins, biocolours and bioactive compounds from agro-produce and by-products: Extraction technologies of oils, oleoresins, bio-colours and bioactive compounds such as

super critical extraction would be developed for value addition to agro-produce and byproducts in the country.

- Technology for cholesterol free milk and milk products: Development of technology for diversification of dairy products into health foods through removal of cholesterol from milk.
- By-products utilization: Crop residues and agro-processing byproducts need to be utilized for obtaining food, feed, pharmaceutical, nutraceuticals, cosmetics and pesticidal products. Technologies for conversion of crop residues and byproducts into value-added products will be developed.
- Modified Atmosphere Packaging and Storage of Perishable Food Products: Post harvest management of perishable food commodities such as fruits, vegetables, meat, fish, etc, require suitable packaging and storage to maintain the quality until it is either consumed or processed. Development of technologies for Modified Atmosphere Packaging and storage for Indian commodities and conditions will be undertaken.
- Biofuels: Some of the agricultural residues, processing byproducts and non-edible oil seeds could be converted into gaseous and liquid fuels for subsequent utilization in agricultural and allied activities. Efficient technologies for conversion of agricultural biomass into fuel-alcohols would be developed.

In the above priority areas jointly identified for research, it has been agreed that in the first instance, training of the Indian participants - about 2-3 scientists per programme (up to six months) in the 7 research programmes would be undertaken (Funding for this would be as provided in HRD component). Specific training details will be drawn up by June 2006.

This would be a prelude to the implementation of actual research programme in the identified areas - to be undertaken over the next three years

Biotechnology

India and the United States agree that biotechnological tools can make an important contribution to the sustenance of the 'Evergreen Revolution'. We share a common goal of translating the results of research in the lab into beneficial products that are delivered to farmers. Subject to funding from the U.S. and Indian governments and bearing in mind possible private sector engagement, focus will be on the following areas:

Transgenic Crops

- Harnessing the benefits of biotechnology: A workshop will be jointly hosted in 2007, involving all stakeholders, for promoting application of biotechnological tools to solve important agricultural constraints, addressing the continuum from molecular research to applied product development and commercialization, with a focus on delivering benefits to farmers.
- Development of transgenic crops:
- Resistance to viruses of economic importance: Initial focus will be to
  - (i) expand existing collaborations begun under the Joint Working Group on Agricultural Biotechnology (e.g. resistance to Peanut bud necrosis virus, Iris yellow spot virus, Tobacco streak ila virus, Tomato leaf curl virus) and
  - (ii) include crops identified in the Indian proposal, such as banana, papaya and cassava. Postdoctoral visits and faculty exchanges to be arranged subject to availability of funds, beginning in 2006 with at least one six-month in-service training visit.
- Tolerance to drought, heat and salinity. Specific crops and genes to be identified during a planning session as part of the "Harnessing benefits" workshop.
- Micronutrient uptake. Specific projects will be identified during a planning session of the "Harnessing benefits" workshop, including iron and zinc utilization efficiency in rice and wheat.

Genomics

India's proposal on pigeon pea genomics will be considered at an appropriate time subject to availability of funds to be discussed further at the next Board meeting.

Molecular Breeding

India's proposal to develop molecular breeding tools in crops and animals will be taken into consideration subject to the availability of funds. Projects could build on existing collaborations, e.g. in legumes and oilseeds. This will be discussed further at the next Board meeting.

Diagnostics and Vaccines

Development of diagnostics and vaccines against selected animal, poultry and fish diseases, such as Avian influenza, Bluetongue, Newcastle Disease, Classical Swine Fever, Neosporosis and Coccidiosis. Collaborative research activities will be developed subject to availability of funds, beginning with six-month in-service training in 2006 in the area of avian influenza. Explore opportunities for private sector engagement.

Additional Training Activities

(i) Food safety analysis in biotechnology for toxicity and allergenicity analysis, faculty exchange (beginning 2006 one U.S. faculty to visit India, two Indian faculty to visit U.S., 2-12 week visits determined by availability of funding and specific training needs of faculty).

(ii) Biosafety of transgenic crops, group training undertaken in India during visits by U.S. faculty (beginning in 2006/07 with 2 U.S. visitors).

(iii) Licensing and Intellectual Property Rights protection: Long-term (3-6 months) and short-term (2 – 8 weeks) training in the process of licensing biotechnology discoveries, hosted by U.S. universities and/or agricultural research institutes and licensing offices (projects to be developed in 2006 for 4 visitors, training beginning 2007).

Water Management

The Board agrees that improvement of water quality and water use efficiency will be vital to the continued growth and productivity of the agricultural sector of both India and USA. The Board agreed to cooperate on capacity building and joint research activities to develop improved technologies and management practices in a framework that incorporates the needs of multiple stakeholders from lab to farm. The following have been agreed upon:

1. Water Quality Management and Remediation

The following activities will be undertaken:

(i) Point and non-point source pollution

(ii) Recycling and re-use of waste waters

(iii) Productive use of marginal and poor quality waters including aquaculture

(iv) Shallow aquifer remediation

2. Assessment and management of agricultural drought

The following activities will be undertaken:

(i) Early warning systems

(ii) Crop condition assessment techniques

(iii) Impact of climate change on droughts and identification of hot spots

(iv) Improved conservation agricultural techniques

3. Soil- Water- Plant Interactions

The following activities will be undertaken:

(i) Water-integrated-tillage interactions on crop productivity and soil and water quality

(ii) Carbon dynamics including sequestration and nutrient pools and fluxes in relation to investment in energy and inputs

(iii) Modelling the impact of water-nutrients-tillage interactions on crop growth and sustainability parameters

4. Sustainable use of Ground Water Resources

The following activities will be undertaken:

(i) Coastal aquifer management – creation of sub surface fresh water barrier for arresting salinity ingress

(ii) Management of Aquifer recharge in water deficit areas

(iii) Deep aquifers utilization and management

5. Use of Modern tools in water management

The following activities will be undertaken:

(i) Integrating space and information technology with water management for planning and decision making

(ii) Decision support systems for management of uncertainty and risk in irrigated commands and rainfed agriculture

(iii) Cost effective water and energy efficient irrigation systems and automation

6. Capacity Building

(i) Water quality management and remediation

(ii) Water use management and policy making

(iii) Watershed management and rehabilitation

(iv) Soil -Water -Plant Interactions

(v) Sustainable use of Ground Water Resources

(vi) Use of Modern tools in water management

7. Implementation Strategies

(i) Develop bilateral research agreement and implement a competitive grant programme

(ii) Facilitate exchange of Professors through sabbatical arrangement

(iii) Facilitate graduate student training by creating enabling environment for their entry and research work

(iv) Provide seed grant for initiation of appropriate research project which would eventually lead to major research programme

(v) Partner with private sector to conduct mutually beneficial research

(vi) Develop training programmes for extension workers (India/US)

8. Action Plan

(i) Identification of US universities/institutions: April 2006

(ii) Indo-US joint workshop in May 2006 for finalizing the technical programme and funding

(iii) Initiation of short term programs from July 2006

(iv) Items at 4 and 5 above will be undertaken subject to the availability of funds

CONCLUSION

The Board enthusiastically agreed to the above Work Plan focusing on capacity building in research, education and extension and forging public-private partnerships that will significantly re-invigorate the US-Indian partnership in agriculture.

The Board believes that the Knowledge Initiative will support the envisioned 'Evergreen Revolution' based upon sustainable, need-based, demand-driven, market oriented agriculture towards increasing rural prosperity.

In reviving and further strengthening our partnership, the Board acknowledges the importance of robust market institutions, efficient, effective and relevant technologies and an enabling environment for investment in agri-business.

The Board agreed to have a digital dialogue immediately after the forthcoming visit of the US President to India in March and later meet in Washington in 2006.

6

Agricultural Extension and Education in India

INTRODUCTION

Extension and education are critical components in development across a wide range of issues, including natural resource management, agricultural productivity, health and sanitation, and watershed management. In order for rural development to be successful, it must be approached from a holistic point of view, and each one of these issues represents an essential part of integrated rural development. Multifunctional extension programs are useful strategies for accomplishing integrated rural development, and can include environmental, economic, health, and agricultural productivity goals. Our paper is designed to discuss how Indian extension can better accommodate a variety of rural development strategies with agriculture at the core.

Specifically we discuss agroforestry, nutrient management in soil conservation, watershed management, and extension programs focusing on linking agriculture and health. In all of our discussions there is an implicit understanding that the technical aspects of extension and training must be married to the socio-economic characteristics of communities where extension agents are working. For some of our papers this understanding is an essential facet of our discussions and the subject of our analysis and critique of extension programs in India. Much of our analysis focuses on India's dry-land regions, where low levels of agricultural production and high levels of environmental degradation are inextricably tired to rampant poverty and malnutrition. Thus, an integrated rural development package must be adapted to local conditions in order to effectively address the multiple and interlocking barriers currently facing the rural populations living in these regions.

IMPROVEMENTS IN AGROFORESTRY EXTENSION IN INDIA

Agroforestry has the potential to significantly improve the livelihoods, economic viability, and agricultural production of small farmers in traditionally marginalized areas. In addition, agroforestry can be used as a tool for changing attitudes towards natural resource management (Glendinning et al. 2001). In order to understand how extension can better accommodate natural resource management through the successful introduction of agroforestry and tree planting, extension agents, both public and private, must strive to understand the socio-economics of agroforestry and of the communities within which they work ("On farm...").

While agroforestry has traditionally been successful in humid and semi-humid areas, it is most needed in dry-land areas, which are characterized by acute climactic challenges and extremely low levels of agricultural production ("Agroforestry Systems in India" 2004 and "On farm..."). The limits to the adoption of agroforestry are many, but must be viewed beyond technical and biotic constraints of the dry-land agricultural regions; the ability of extension agents to recognize socio-economic characteristics of farmers is essential in learning how to target potential innovators who can then generate change within their communities (Mahapatra and Mitchell 2001).

More emphasis on the development of participatory programs, more efficient targeting, and intensified focus on the traditional communication channels will increase the adoptability of agroforestry in dry-land communities. In turn, an implicit understanding that the attitudes of farmers towards natural resource management are based on agroforestry successes and failures will help extension to increase the environmental reclamation of degraded landscapes, especially those in dry, rain-fed agro-ecosystems.

Dry-land agricultural areas in India face specific problems that must be overcome for both improved agricultural production and environmental conservation. Agroforestry can help overcome limited water availability, fuel shortages, degraded soils, and a lack of biodiversity ("Agroforestry Systems in India" 2004 and CRIDA). When trees are used that are adapted to water-limited areas, such as *Gliricidia*, communities can realize both agricultural and environmental benefits (CRIDA). In general agroforestry produces a multitude of benefits, including increased income from selling surplus goods, improved environments and enhanced biodiversity, increased access to education through income generation, shelter from harvesting trees for building materials, improved health when medicinal plants are incorporated, a greater diversity of food sources, and an increase in the energy of an agricultural system because manure can be used for fuel instead of food ("Agroforestry Systems in India" 2004).

Specifically in rain fed agro-ecosystems, like those found in dry-land agricultural areas of southern India, successful agroforestry projects can increase the organic matter in soil, improve the physical properties, increase crop yields, increase the availability of nitrogen in the soil, increase the water-holding capacity of soil, and reduce soil erosion. In addition, there is a reduction in pollution because of a decrease in chemical fertilizers. Most environmentally significant, the biodiversity of an agricultural area is enhanced (CRIDA). Agroforestry's scientific benefits as listed above have been well documented within the humid and semi-humid agricultural regions, and the challenge for extension agents is to successfully spread these benefits in dry-land areas where, traditionally, farmers have been marginalized due to resource limitations (such as soil-fertility and water availability).

These benefits are of utmost importance in creating a sustainable agro-ecosystem in dry-land agricultural areas, and they cannot be realized without appropriate extension and education. Most importantly such extension and education must be localized and participatory in nature (Glendinning et al. 2001 and "On farm..."). In order for training and education to be localized, extension agents must understand the important role of socio-economic factors as well as the technical aspects of agroforestry, and therefore they must begin to focus more on the social and communication constraints of the specific villages and communities within which they work (Mahapatra and Mitchell 2001). Indeed the foundation of success is the "participation of the people for whom the forestry programme is intended" (Mahapatra and Mitchell 2001).

This participation of local residents will help transform the extension programs from the current "top down" structure, pioneered in "transfer of technology" methodology, to "bottom up" participant defined priorities and goals (Mahapatra and Mitchell 2001). In this way, participants are much more responsive to the education and training offered by extension agents (Glendinning et al. 2001 and "On farm..."). In fact, the potential participatory nature of agroforestry programs in rural communities can be a useful vector for knowledge generation that incorporates local information and global science (Brodt 1999). Villagers and communities, depending on a number of variables, are continually gathering knowledge from various sources, and extension agents should do less to emphasize the strict dichotomy of global science and indigenous knowledge as it can be harmful to the adoption rates of tree management and agroforestry (Brodt 1999 and Mahapatra and Mitchell 2001).

Such a dichotomy will ultimately only deepen the impression of 'otherness' of indigenous and rural people's knowledge and thereby reinforce the developed/ underdeveloped polarization that many of these same researchers and development experts are trying to overcome...Rather than solely emphasizing differences, scholars might accomplish more by seeking out points of congruence and interface between local realities and larger-level systems (Brodt, p.362, 1999).

Ultimately, Brodt and others advocate that synergy instead of trade-offs be emphasized when extension agents use a variety of knowledge sources to advocate certain practices. Communities within dry-land regions of India have already developed knowledge concerning tree management; however, these knowledge systems are "open", thereby allowing an exchange of information which can be more conducive to innovation than enhancing a "closed system" through strict juxtaposition of local knowledge with global science (Brodt 1999). By allowing the participants to innovate beyond their previous knowledge capacities, extension agents are facilitating local ownership of knowledge, which enhances the success of programs and promotes long term commitment on the part of the farmer and community.

Participant defined priorities and goals and an incorporation of local knowledge by extension agents must also be complimented by designing programs that better target potential users, early adopters, and innovators. Programs must be designed to address different family units, income levels, ages, land tenancy, farming experience, castes, occupations, urban contact, degrees of social participation, farm modernity, household assets, access to capital, and varying off farm incomes (Mahapatra and Mitchell 2001). Methods have been developed that, when using these and other variables, can successfully identify those farmers that will be "tree planters" within a subsistence farming community (Mahapatra and Mitchell 2001). It is essential that, before an extension program is implemented, "the beneficial characteristics that might promote or impede local involvement in forestry programmes" are defined (Mahapatra and Mitchell 2001). Mahapatra and Mitchell have identified several key factors that make a farmer into a potential "tree planter" including "progressive attitude", "membership in village organizations", "wealth status", and "perceived risk concerning agricultural production." In addition to such characteristics, they also note that off farm income generation is generally very complimentary to the low labor inputs of agroforestry (or home gardens).

Most importantly, it has been shown that social or economic factors cannot account for the adoptability of tree planting and agroforestry; these factors must be combined in order to successfully target farmers who will plant trees. Another important note is that land area and tenancy are important, but mistakenly overemphasized when targeting farmers: "Classification of farmers by size of landholding, as [is] usually done by agricultural extension workers in India, is not suitable for farm forestry extension" (Mahapatra and Mitchell 2001). This is especially important for agroforestry extension in dry-land areas where farmers are operating on very small plots of land—further supporting the spread of agroforestry beyond humid and semi-humid areas. As discussed before, an exchange of information and knowledge generation are both critical in innovation and the success of agroforestry, and the model developed by Mahapatra and Mitchell also emphasizes this critical point by observing that membership in village organizations and contact with urban populations are essential indicators in targeting adopters and innovators.

By better understanding how to target specific farm types, farmers, and communities extension programs can shift away from the "conventional approach" of delivering standardized tree seedlings and vague management advice to all farmers. In this way they can concentrate their energy, time, and resources on specific farm types and farmers as well as identify the "uncertain" group that could either adopt or dismiss agroforestry (Mahapatra and Mitchell 2001). By helping change the attitudes of "fence riders" and targeting the farmers that are very willing to innovate, agroforestry programs will see much more long term success and increase the adoptability of tree planting and, thus, natural resource conservation.

The participatory model of extension and education advocated throughout this paper as well as the emphasis on targeting potential adopters and innovators is underscored by a better understanding of local knowledge dissemination and technology transfer among rural farmers in India's dry –land regions. When extension agents invest time in learning how villages and farmers communicate, they also see that participatory programs and targeting initiatives are critical in establishing successful agroforestry farms; "the decision to adopt agroforestry was found to be determined by the farmers' attitude to agroforestry, which in turn was shaped by information received through farmer-to-farmer and farmer-to-extension contact. The mode of communication was important and, to be effective, needs to be customized for each target group" (Glendinning et al. 2001). Glendinning has found that the most important channels of communication for farmers, when innovating and learning, are extension agent, neighboring farmers, and group meetings.

These channels of communication indicate that participatory programs and better targeting by extension agents are appropriate ways to disseminate knowledge and change attitudes. Agents can target potential innovators who are capable of bearing more risk than other farmers. When these innovators generate both economic and environmental benefits through agroforestry, it is more easily communicated to neighboring farmers and members of group organizations. Neighboring farmers are more likely to adopt agroforestry if they see it has been successfully done by members of their own community. Successful demonstrations by villagers and other farmers decrease the perceived economic risk of agroforestry. "In regions where smallholders predominate [like dry-land areas in India], promoting change within the existing farming system may therefore require substantial diversion of resources in order to develop effective communication systems, skills, and media" (Glendinning et al. 2001).

It is therefore equally important for extension agents to assume the role of facilitators when first entering communities; they can create or enhance pre-existing village meetings and farmer-to-farmer contact. Extension agents can take advantage of the local social structures to increase the "adoptability" of agroforestry. In addition for small landholders personal communication with neighbors and extension agents are the two modes of communication and knowledge sharing that are most strongly linked to potential adoption and tree planting (Glendinning et al. 2001). Understanding the communication channels and constraints will enable extension agents to enhance village communication and capacity building of farmers, which will, in turn, help demonstrate the low economic risk of agro forestry as well as the environmental benefits and improvements in agricultural yields.

Local knowledge, participation, and better targeting by extension agents are critical in establishing a long term commitment to agroforestry in dry-land regions of India. The importance of successful agroforestry extends beyond satisfying agricultural production and economic goals, it is also critical in establishing an understanding of why natural resource management and environmental conservation are important in the sustained productivity of ecosystems. The future of natural resource management depends on the attitudes of farmers which are distinctly shaped by the information they receive from communication channels guided by extension services. Extension agents in India have established the scientific base and technical background to support agroforestry adoption; however they have yet to master the socio-economic tools to analyze communities and enhance local social structures that enable the adoption of agroforestry and facilitate knowledge exchange and generation, critical components of successful agro-ecosystems.

If extension agents can use socio-economic analysis to better understand the local conditions in which they operate, the adoption rates of agroforestry will increase, and the overall environmental benefits will be realized. These environmental benefits are critical to sustaining agricultural production to feed a burgeoning population. One of these, nutrient management in soil fertility, is of overwhelming importance for cropping systems. Further examination of nutrient management in Indian dry-land agriculture is needed to understand the future of agricultural productivity, and the overwhelming importance of agroforestry in degraded landscapes. Such novel approaches aimed at better understanding local conditions and farmer capacities and attitudes toward adopting improved land management are also a useful tool for analyzing nutrient management issues in much of India's degraded dry-land agricultural soils. As we will see in the next section of our group paper, agro-ecological constraints to enhanced food production often are exacerbated by poor land management practices among resource poor farmers living in these areas. Thus, new soil and cropping systems management must be adapted to local conditions in order to ensure sustained agricultural production without depleting the existing natural resource base.

SOIL FERTILITY MANAGEMENT IN DRY-LAND INDIA

Expensive soil and water management technologies, high capital and energy requirements and "top-down" development approaches have not alleviated soil fertility problems in Indian dry-land agriculture. Rather, in spite of many costly investments in agricultural extension, soil degradation worsens and production remains extremely low. On the technology side, the majority of marginal and small-scale farmers throughout the region are ill equipped to benefit from new farming technologies that often promote the excessive use of agrochemicals and other external farm inputs.

Thus, future thrusts in agricultural extension should focus on promoting more sustainable alternatives to India's ongoing downward spiral of declining soil fertility and low crop yields. One effective tool for analyzing such approaches on both a theoretical and practical level is to learn from the vast body of available indigenous knowledge on soil fertility management. Traditional knowledge and practices related to dry-land agricultural production have been adopted and modified by Indian farmers based on local needs and resource availability. Advances in dry-land agricultural management must seek to build upon- rather than replace- indigenous soil fertility management practices in a cost effective and sustainable manner that allows for both enhanced agricultural production levels and improved natural resource management.

Most of India's dry-land regions consist of two primary soil orders- Vertisols and Alfisols- both of which pose significant problems for agricultural management. Vertisols, with their characteristic shrink-swell behavior of clays as they go through wetting and drying cycles, have low structural stability and thus, are difficult to manage. Due to the critical timing of performing tillage operations on these soils, some farmers have dubbed them "24-hour soils" for their tendency to be too hard and dry to cultivate one day and too wet and sticky the next (FAO website, accessed 3/05). Alfisols, meanwhile, are characterized by very low natural levels of soil organic matter (SOM) and nutrient availability. They also have poor water holding capacity and are prone to extremely high rates of surface runoff and erosion when left bare under heavy rainfall conditions, such as those seen during India's monsoon season (ICRISAT 1984). Thus, agricultural production on these soils is extremely low and land degradation very high.

Traditionally, agricultural extension in dry-land India has focused on improving overall production levels by eliminating or reducing the effects of a number of agroecological factors. In dry-land areas in India and elsewhere, lack of adequate water supplies and poor water management are the most prominent limiting factors for improving agricultural productivity. Public extension in India's dry-land have thusly invested many resources into high input irrigation systems and rainfall collecting and harvesting technologies. Comparable investments have been made to promote and disseminate new high yielding seed varieties (HYVs) of basic grain crops and cereals, such as rice, wheat and maize. HYVs have fared very well under favorable on-farm conditions with the application of external inputs, but have not been as effective for resource poor farmers in India's dry-land regions. Much of this is due to the lack of available land, labor and capital needed for purchasing and effectively managing the technologies.

Poor land management practices also have significantly contributed to the concomitant problems of loss of soil fertility and low agricultural yields. Deforestation and overgrazing are the two primary factors contributing to soil degradation, and crop yields are further limited by mismanaging tillage operations and fertilizer applications as well as mistiming planting dates for optimal harvest (Hobbs 2005). Research has shown that up to 40 tonnes per hectare of valuable topsoil may be lost following 15 years of cultivation (Magdoff and van Es 2000). Much of this loss may be attributed to intensive tillage operations that leave the soil bare and unprotected for long periods of time. As soil organic matter (SOM) is washed away by wind and water, soil fertility is reduced and crop yields subsequently decline. SOM content can be replenished and greatly enhanced by simply not removing crop residues and farmyard manure (FYM) from the system; they are left to decompose *in situ* or incorporated into the soil profile. However, high demands for fuel and animal fodder result in the removal of many valuable nutrients from the agricultural system, and soil fertility remains low.

In spite of some of traditional dry-land agricultural systems' shortcomings, there exists a vast and growing body of knowledge on indigenous soil fertility management practices. Farmers in Karala and Karnataka have been planting vetiver grass (*Vetiveria zizanoides*) for more than a century as a means of erosion control, and farmers throughout various dry-land regions are reported to have been incorporating animal and green manures into the soil and practicing agroforestry and intercropping (Rajasekaran and Warren 1995). Research shows that such practices are most effective when combined with the judicious use of crop rotations to suppress the prevalence of weed, pest and pathogen problems. Some of these indigenous practices, such as the planting of *Prosophis juliflora* in alkaline soils, have proven highly effective in the reclamation of low fertility, degraded soils throughout India's dry-land regions (Dagar et al. 1995, 2001, 2002).

If Indian dry-land agricultural extension hopes to increase productivity levels without further depleting soil fertility, scientists and extensionists must better understand the important role of indigenous knowledge and soil fertility management practices in achieving such goals. Local "best-practice" technologies have the inherent advantage of being adapted to local agroecological and socio-economic conditions. In contrast, top-down technology transfer (ToT) approaches are insensitive to such important dry-land agricultural variables as rainfall, soil type, vegetative cover, available farm inputs and off-farm nutrient demands, and farmer adoption rates of new technologies remain alarmingly low as a result. Thus, extension must look to complement- rather than displace- indigenous knowledge and local best practice in India's dry-land regions. Doing so increases the likelihood of new technologies being adopted and modified by farmers, and promotes managing local resources more sustainably and productively.

Researchers and extensionists have been experimenting with more participatory approaches in dry-land agricultural production. This change has coincided with increasing farmer use of private, fee-based extension systems in India. Participatory rural appraisal (PRA) is being used to collect, synthesize and analyze available information on indigenous soil fertility management.

This information will be used by extensionists and policymakers to promote resource-efficient agricultural development that is more compatible with the agroecological and socio-economic conditions faced by both large- and small-scale farmers alike. Private, fee-based extension programs have the advantage of direct accountability to the farmers they serve, but lack of money to pay for their services is a problem for many small-scale and marginal farmers. Dry-land agricultural extension, whether public or private, must incorporate scientific knowledge into indigenous knowledge systems and local best practices in order to meet the needs of both better endowed and resource poor farmers throughout the region.

One example of a current agricultural extension initiative that is attempting to incorporate both scientific and local knowledge and practices for managing soil fertility is the current promotion of conservation agriculture (CA). Rice-wheat farmers in the Indo-Gangetic Plains have successfully adopted and modified the no-till package for wheat cultivation, though many farmers continue to use traditional tillage and flooding practices for cultivating rice. CA advocates the permanent use of no-till technologies and practices while also maintaining permanent or semi-permanent soil cover and using regular crop rotations. Not surprisingly, researchers and extensionists are finding no-till adoption most successful in regions already having a strong tradition of growing green manure cover crops (GMCCs) and adding farmyard manure to enhance soil fertility (Lal 2004). These factors are strongly influenced by several socio-economic factors such as high demands to feed crop residues to animals and use manure for fuel.

History demonstrates that new technologies relying on high capital and energy requirements are unlikely to significantly impact resource-poor farmers and their families. Examples of traditional soil fertility management in India portray a wide variety of local best practices, many of which are capable of increasing on-farm productivity without depleting or otherwise damaging natural resource bases. Similarly, public and private agricultural extension systems must seek to promote an effective balance between improving on-farm productivity and sustainable resource management. Furthermore, the technologies must be available and practical for farmers of all sizes and resource allowances. In some dry-land regions of India, conservation agriculture practices such as the use of no-till, green manure cover crops and regular crop rotations have been successfully adopted by both large- and small-scale farmers to help alleviate soil fertility and production problems. Indian farmers that practice no-till and other components of conservation agriculture have reported decreased water, fertilizer and labor requirements and increased levels of production. The resulting increased profitability for Indian dry-land farmers is the primary cause for widespread adoption.

By increasing agricultural productivity and profitability while also enhancing soil fertility, conservation agriculture practices have tremendous potential to reverse the downward spiral of declining soil fertility and low crop yields. They are also highly capable of being adapted to local needs and levels of resource availability. Some of the technologies also require enhanced levels of knowledge and training, and extension services would be responsible for ensuring that new information and materials are made locally available and practical for dry-land farmers. Researchers and extensionists must also seek to improve no-till rice production, and introduce new cover crop species that produce enough biomass to provide for soil cover and animal fodder in dry-land silvopastoral systems. Such advances are likely to be adopted most readily when researchers and extensionists use participatory methods to involve local farmers and build upon preexisting indigenous knowledge systems and practices.

Yet for soil fertility management to be effective in the long run, it must be part of an integrated community natural resource management program that includes laws prohibiting the clearing of forests and the burning of crop residues as well as a plan for community watershed management. Limited water supplies and poor water management are the two primary limitations to agricultural productivity in India's dry-land soils. The next section of our group paper will discuss what is needed for establishing and supporting integrated community watershed management programs in these dry-land regions.

WATERSHED MANAGEMENT FOR DRY LAND AGRICULTURE IN INDIA

During the past five decades of planned development, India has achieved spectacular increase in food grain production. Green revaluation enabled us to enter in to a new era in input use. The use of high yielding variety seeds, fertilizers, plant protection chemicals have stepped up manifold. Development of major and minor irrigation projects had gone up. But still the over all gains had not been evitable shared by the farming community. This is mainly because two thirds of agriculture in our country ifs rainfed.

According to water for irrigation would emerge as a critical bottleneck. The requirements of groundwater, it will lead to over exploitation of ground water in several status and by 2020 as much as 71% of irrigation area is likely to be from ground water sources. Closely related to this is the problem of degradation of land in the rainfed areas.

Rain fed agriculture in India is characterized by low productivity, degraded natural resources and wide spread poverty. Most of millions of people living in our country depend on agriculture and natural resource management for their livelihoods, so development planners are eager to implement productive, environmentally sustainable land and water management. It is in this concept of watershed development has been introduced in our country.

Watershed

It could be defined as an area from which the runoff drains through particular point in the drainage system.

Main aims of the watershed management programme in rainfed areas are:

- Enhancing crop production and increasing farmers' income.
- Minimizing degradation of natural resource base.
- Principles of watershed management:
- Utilizing the land according to its capability.
- Adequate vegetative cover during the rainy season.
- Conserving as much rainwater as at the place where it falls.
- Effective utilization of surface and ground water resources.
- Adopting integrated soil, water, nutrient, and pest management options.
- Increasing cropping intensity and land equivalent ratio through intercropping and sequential cropping.
- Safe utilization of marginal lands through alternative land use pattern.

Water Management Structures: Waterways

Waterways are made between fields to safely dispose off excess rainwater out of the field without much soil erosion.

Farm Ponds

Grassed waterways lead to farm ponds where excess rainwater is collected for supplemental irrigation during the dry periods or for animal use.

Check-dams and Gully Plugging

Check-dams and gully plugging control soil erosion as well as facilitate percolation of rainwater into the ground.

Percolation Tanks

Percolation tanks such as check –dams and gully plugging help in reaching the groundwater.

In our tour programme we were visited Kothapalli village in Andrapredish.it is organized by the international crop research institute for the semi-arid tropics ((ICRISAT) has launched a scheme on watershed management. Which offers a pivotal linkage between the farmers needs resources and local environment and an institute. Under this condition ICRISAT identified Kothapalli village and established watershed at 2000. The over all aim is to combine the generalized information to suit profitably to individual farming situations. Different department people joined together and maintain the watershed in Kothapalli village.

Rain Fall and Soil Type: The climatic condition of this village is hot, semiarid region with average annual rainfall is 500-600mm.Most of the annual rainfall is received during the June-Oct. 80% of rainfall is received during the summer season. Off-season rain is used for preparation of land.

In Kothapalli village the soils are widely distributed. The major soil types are black soil and incept sols. Maximum rainwater loss through evapotranspiration and runoff because the very low moisture holding capacity of the soil.

Cropping Pattern: In Kothapalli village-cropping pattern has changed, before the establishment of watershed they were cultivated only one crop, now they are cultivating three crops.

- First crop- Paddy.
- Second crop- Sorghum, Safflower, Sunflower, Pigeon pea.
- Third crop- Vegetables.

It shows management the effective utilization of land.

Benefits of proper watershed management

* Increased crop productivity and profits.
* Reduced soil erosion and better land quality

* Improved surface drainage and reduced water logging.
* Increased ground water recharge
* Increased water availability for drinking for humans and animals.
* Long-term sustainability of agriculture.
* Improved agro-ecology of watersheds.

HUMAN DEVELOPMENT THROUGH INTEGRATED HEALTH AND AGRICULTURE EXTENSION IN DRY-LAND INDIA

The previous sections have dealt explicitly with agricultural management systems in India and demonstrated the importance of considering multi-strategy solutions in agriculture development. The severe environmental constraints in dry land agriculture merit a dynamic approach to agriculture improvement that encompasses not only nutrient management and agro-forestry, but particular attention to local watersheds. Though governmental extension services in India have adopted agroforestry and some watershed management programs in their extension training, little emphasis has been placed on human and environmental development where decisions and action plans are decided at the community level. Public extension in India does not have the resources to implement integrated programs that value the participants' development even more than the farm systems' development. In response, private extension, primarily in the form of non-governmental organizations (NGOs), has begun to expand beyond the reaches of the production-oriented public sector, both in regional coverage and methodology. Integrated conservation and population programs are based on the principle that human health and development is essential to the sustainability and the protection of ecosystems, natural resources, and communities. An analysis of the global relevance, the principles of human and integrated development, their advantages, and a case study of community based population and environment program[1] elucidates the magnitude of change integrated programs can generate.

By nature, communities are the interface between humans and the environment. At this interface there are numerous factors, including population, socio-economic, political, biophysical, and cultural components that influence the lifestyles of community members and their interactions with the environment. Historically, India's policies, aid programs, and extension have dealt with these development issues through separate ministries, rather than approaching them from the viewpoint that people and the environment are interwoven. Globally, the existence of this linkage between human health and the environment was acknowledged at the United Nations Conference on Environment and Development in 1992. The primary health care and environment millennium goals determined at this conference clearly state the international community's acceptance of this linkage between the environment and human health. Since then, efforts have been made to discover, design, and/or analyze initiatives and programs that combine community-based population and environment.

Though the global community and perhaps the Indian government have acknowledged the logic of combining health and environmental services, both are far from implementing such programs. Limited public resources coupled with burgeoning problems and dogmatic development strategies prevent more adoption of integrated community development programs in India. The private sector has started to fill this neglected area, focusing on human development as an essential step towards sustainable and scaleable programs. The inherently interwoven framework of communities and human interactions with the environment attest that agriculture development is limited without human development. Though agriculture development may improve the farmer's production, increase his income, and preserve the land for future generations, the sustainability of these results and the expansion to other individuals and communities may be compromised if health and socio-economic conditions are ignored and the people themselves are not taught self-reliance.

There are several conceptualizations and methods of achieving human development that vary between organizations, but most share root convictions and the final goal is always self-reliant, confident individuals and communities capable of making informed choices and decisions. According to Don Jose Elias Sanchez, whose work with poor farmers in Honduras is documented in the *The Human Farm,* human development grows from the conviction that only strong hearts and heads can guide the hands. Sanchez's training farm is centered around exercises and sessions designed to help farmers realize the strengths and understanding they bring to agriculture (Smith, 1994). This ground-up, holistic approach develops human capacity in a working context that farmers can transfer back to their own communities. UNICEF on the other hand, perceives human development in three basic principles whose achievement varies across place. The first principle is "sustainable human development evolves from a self-reliant understanding of local needs and resources (Taylor-Ide and Taylor 1995)."

Communities have an intuitive sense of the dynamics of the local ecosystem, economic potentials, market forces, and social development variables (Taylor-Ide and Taylor 1995). Recognizing these dynamics and giving community members the skills to collect data, respond to problems, and analyze situations improves self-esteem and can lead to a more sustainable project. The second human development principle is "action must grow from a combination of a bottom-up and top-down programming (Taylor-Ide and Taylor 1995)." The informed decision-making needs to be based on a three-way inclusive dialogue between the *entire* community, government and agency officials, and experts. The third principle is "sustainability is only possible when action grows from community participation and self-reliance (Taylor-Ide and Taylor 1995)."

The success of an agriculture development project is contingent on the community's ability to analyze and respond to new problems after the withdrawal of the development agency. If communities are not taught the skills to understand and evaluate their own environment, the likelihood of expansion and/or sustainability is diminished. Here, in the third principle, is where the ideas of *The Human Farm,* UNICEF, and most human development schemes converge: the development of humans themselves empowers the individual and community, enabling the expansion and growth of their potential through their self-reliance and independence. This form of development is key to the future success of extension in India; the investment of time and training in people would assist the government in addressing the multitude of problems rural Indians encounter. The case study presented later in the paper elaborates on human development in the Indian context.

The integration of health and environment is advantageous to community-based extension because it addresses community interests as a whole, rather than as separate problems that are exclusively independent.[2] Two of the foremost issues that need to be addressed in rural India are public health and agriculture. The leading causes of death in India include neo-natal, diarrhea, and pneumonia; infections and diseases that are all exacerbated by malnutrition. Anemia, a micro-deficiency caused by lack of iron, is prevalent in more than 69% of pregnant women in India (CDC, 2000). Vegetarian diets, lack of protein and bio-available iron sources contribute to a deficiency that has severe consequences such as poor brain development in children, compromised immunity, maternal mortality, and reduced work capacity (Stoltzfus, 2005).

The majority of these deaths and illnesses could be prevented with proper health interventions and food supply improvements. The health of people is not contingent only on the availability of food, but the sanitary conditions and practices of families. Primary health care such as access to rural heath posts, education on proper infant and childcare, maternity care, reproductive care, portable water, and sanitation are all basic needs and rights the greater part of rural Indian people are denied. Beyond the logical sentiment that this is iniquitous, researchers at the forefront of nutrition and economics argue child and adult malnutrition reduces labor output, GDP, cognitive ability, and motivation (Pinstrup-Andersen, 2005). It is not only beneficial to improve the health standards of people for their own wellbeing and ethics, but for the wellbeing of the country.

Agriculture and public health are linked in two ways. First, without sustainable farming techniques, diversified crops that bring in multiple nutrients, and income for supplemental foods, the consistent yield, sustained nutrition and capital of the farmer and his or her family are compromised. Second, poor health, whether in the form of micronutrient deficiencies or parasites, can contribute to an inability to address farm issues with the energy and vigor expected from a healthy farmer and family.

This linkage of health and motivation is not only an example of the significance of linking health and agriculture, but it further validates the need for health in human development programs. Linking health and agriculture has the advantage of addressing these principal issues at the same time. Furthermore, while it may be incongruent for experts or professionals to tackle health and agriculture together because they cross disciplines, it is natural for rural communities to perceive both as interconnected and integral to their personal growth. Communities are affected by multiple factors, and a holistic perspective is intuitive to them. Advantages of holistic approaches to community development reach further beyond the more evident reasons discussed.

This health and environment approach to community development can facilitate entry into communities. Projects that linked family planning and environment often found barriers were more easily overcome when environmental services provided the means of introducing family planning to the community (Engelman, 1999). Reproductive health is a potentially sensitive topic in India. These programs can generate confidence and community acceptance, in addition to being a way of creatively linking reproductive health to agriculture thus reducing potential hazards of discussing family planning. Environmental metaphors may include comparisons between the importance of spacing crops and the spacing of children (Zelaya, 1999) or the need to prepare for children just as the land needs to be prepared for seeds. Additionally, the link has provided a gateway for women into the environment sector and men into the reproductive health arena.

Community-based population and environment programs can meet the needs of remote communities more efficiently than most sectoral (Engelman 1999), public sector, and advanced technology driven programs. The people that need health and environmental services the most are those living in remote areas with little access to roads, schools or health services. These areas are usually the last to receive government services, the least likely to accept such services when they are available, and the slowest to change attitudes and behaviors (Vogel and Engelman, 1999). As a result, these areas in particular require strong community based holistic approaches to meet varied needs and improve adoption through the establishment of trust and the thorough understanding of community culture. Integrated and human development programs have the potential to actually generate human capital in these resource-limited regions.

This development of people is exceptionally important, as it is unlikely that further assistance will be readily accessible on the completion and withdrawal of the project. The strong base of combined health and environment programs gives the organizations the latitude necessary to establish this trust, understand the communities, and improve self-reliance in these remote areas. There are other advantages to integrated development programs, including the inclusion of men and women in typically gender sensitive topics, cost effectiveness, and sustainability. The case study will demonstrate these advantages and those already discussed. The following section profiles the work of World Neighbors, an NGO presently working in India that has approached development holistically with an emphasis on health, environment, and specifically human development.

World Neighbors (WN) works in Karnataka, a dry land state in southern India. Though the state is considered at the median level of development in major sectors, the Human Development Report 1999 "describes the regions covered by WN programs as 'areas of darkness' with HDI indicators that fall below the sub Saharan African countries (WN, 2005)." The region is typical of many dry land areas around the country, with agriculture depending on scanty and erratic rains and over 90% lacking irrigation (WN, 2005). As a result, farmers and shepherds of the region often migrate in search of food, fodder, and employment (WN, 2005). Malnutrition and infant mortality rates are high in comparison to other areas in the state (WN, 2005). In 1998, WN began working in reproductive health in two villages; Nellur and Yelenavadgi, where agricultural activities focused on dry land agriculture, sheep health, and wool processing were already ongoing. Over the past three years WN has partnered with 8 additional NGOs to facilitate activities and trainings in order to meet several objectives:

1. To organize target groups to undertake development activities
2. To create employment opportunities for the rural poor and establish training centers for self employment for youth and women
3. To plan and implement various rural employment and development programs with the cooperation of local groups for the all round development of the people
4. To sustain all development activities targeted to the groups (McKraig, 2002).

As the objectives demonstrate, WN work in Karnataka focuses on human development in order to create sustainable, fruitful ventures, regardless if it's in the agriculture, health, or micro-credit sectors. One particular aspect of the program targets the formation of women's groups through a three-step process. The first phase involves WN and the NGOs holding initial discussions with agriculture groups about the formation of women's groups in the villages (McKraig, 2002). Groups are established and reproductive health workers (RH) are identified and trained (McKraig, 2002). In the second phase, the capacity building of women's groups with the help of the RH worker takes place.

The initial activities of the groups include the selection of leaders and training of leaders in such topics as group management, decision-making, recording keeping and monitoring and evaluation techniques. In the third phase, women's groups conduct activities independently (McKraig, 2002), the RH worker visiting the groups only every three or six months (McKraig, 2002). The first two phases involve the intensive support of WN and the partner NGO to help establish the women's groups, develop capacity, and implement activities (McKraig, 2002).

These steps are essential to the successful development and coordination of the groups. Only when the skills and capacities of the groups are firmly established, about a three-year process, do WN and the NGO withdraw further and the women's groups conduct the activities independently (McKraig, 2002). Though the phases may be lengthy and intense, evaluation of the program demonstrates that women are empowered in many aspects of their lives.

An evaluation of the reproductive health component of WN's work was conducted in 2001, with indicators suggesting the human development methodologies improved the capacity of women to make informed decisions and the integration of health and agriculture was beneficial. Key findings from the evaluation state that "significant changes were reported in key reproductive health practices and rates for service use (McKraig, 2002)," compared to non-intervention villages. Additionally, "there were reported changes in indicators of women's status (McKraig, 2002)." The women indicated that group activities improved decision-making, property ownership, girl child education, and reduction of violence against women (McKraig, 2002)." Women felt they were more confident "to organize and undertake a variety of initiatives ranging from closing bars and gambling houses to using civil disobedience to improve water supplies (McKraig, 2002)."

The linkage between men's agriculture groups and women's groups greatly contributed to the success of the women's groups. The agriculture group members assisted in the formation of the women's groups, encouraged their wives to attend, convinced other men to do the same, and supported them in terms of organization of events and discussions (McKraig, 2002). The linkage also encouraged communication between the men and women about dry land agriculture and "improved awareness of women's group members regarding key agricultural practices (McKraig, 2002)." Another advantage of the program is the integrated model appeared cost-effective (McKraig, 2002). Not only did the program improve the group member's livelihoods, but spread to non-group members. Finally, the inclusion of trainings in evaluation, monitoring, and implementation techniques helped ensure the sustainability of the program. "The women's groups are confident in their ability to sustain key activities after the phase out of BSRDS (McKraig, 2002)." The development of human ability encourages the scaling up of a program; it is a positive step towards the expansion of the program to other communities.

The evaluation of the reproductive component of WN's work in Karnataka demonstrates the value of integrated programs. WN states on their Karnataka webpage, "the most important determining factor of success is the involvement and integration of various community groups (WN, 2005)." The apparent exclusivity of health and agriculture to some professionals is in direct contrast to rural community's instinctive perception that health, agriculture, and a myriad of other issues are innately associated. Furthermore, the case study demonstrates that the development of human capacity enables communities to determine the focus of their projects and their own future. The development of human capital is one of the most efficient methods of any form of development as it empowers people to actually *do* the development themselves and enables communities and individuals to grow and expand beyond the original confines of their poverty. This form of development is part of a viable solution to the multiplicity of problems India faces.

CONCLUSION

"The processes through which past societies have undermined themselves by damaging their environments fall into eight categories: deforestation and habitat destruction, soil problems (erosion, salinization, and soil fertility losses), water management problems, overhunting, overfishing, effects of introduced species on native species, human population growth, and increased per capita impact of people (Diamond 2005)."

This quotation from Jared Diamond's nonfiction book *Collapse: How Societies Choose to Fail or Succeed* reveals the profound impact human's management of the environment and natural resources has on the quality of their future existence. While the societies Diamond describes didn't necessarily disappear or spontaneously combust, their populations either declined or suffered monumental setbacks in their quality of life. Furthermore, the societies did not collapse based solely on one of the eight categories, rather they collapsed based on the compounded affects of a combination of these categories. Four of the eight categories that undermine communities were discussed in this paper: deforestation, soil problems, water management problems, and human population growth. The occurrence of these issues in our paper and in Diamond's book is not a coincidence; the topics are acknowledged global concerns that must be addressed if we hope for an equitable, sustainable future. Our paper demonstrates that multi-functional approaches to environment and human development are integral to successful management in dry land regions of India. Diamond's analysis of the collapse of societies is further evidence that if these issues are not tackled from an all-encompassing perspective, the triumph over one problem may be undermined by the eventual erosion of another. In the dry land region of India, where harsh environmental conditions interact with a growing population and little agricultural support, the reality of the situation merits the development of multi-functional extension programs.

7

ANIMAL AGRICULTURE IN INDIA

Greater number of Indian population live in rural areas and engaged in agriculture. But the growth rate of agriculture is only 1.4 per cent in last two decades whrere as growth rate of West Bengal has remained at 3% for last few years.But over all growth rate of GDP in India about 9%may be considered good in terms of economy.Such low growth rate for gearter number of population engaged in agriculture facing crisis resulting suicide of farmers due to various stress factors related to individual economic crisis specially loan taken but failure to return the debt due to loss of agricultural productivity in many states. Scenario is bit different in the State of West Bengal with increase of purchasing power to the rural population.The success came because of land reform and distribution in legal manner.But the future is bleak if there were no industrialization along with nutritious foods security which has linkage with population health.

Chart showing growth of Agricultural Productivity in Developed(Yellow colour)and Developing Counties(Violet colour).

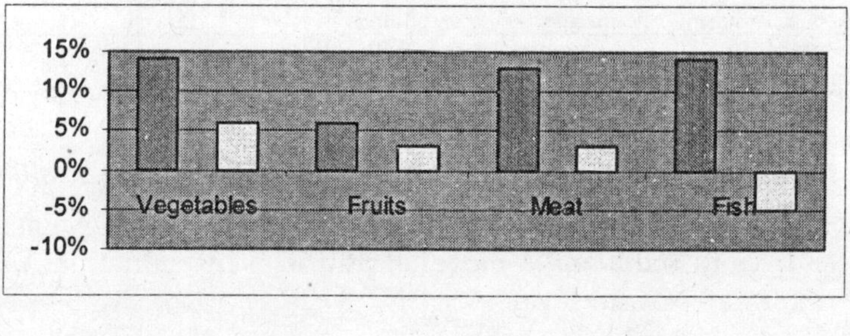

The growth rate of agricultural productivity obviously found to be lower with higher GDP but no food insecurity in developed countries.Unlike developed countries growth rates of vegetables,fruits,Meats and Fishes etc. in India are higher comparatively.

So,it is high time for India to develop Animal agriculture going beyond calories only. The State Budget of West Bengal has enhanced to Rs.15 crore 85 lacs for the year 2006-07 from previous year of Rs.11 crore 96 lacs to speed up the Animal resources productivity for income generation by self-employment.10 lacs Chicks were distributed among 14000 groups which has been decided to enhance 20lacs for the current year.There is a separate new department called the Ministry for self-employment,Govt.of West Bengal.The Ministry works like Gramin Bank of Prof.Yunus in Bangladesh.

In 10th Animal Development week Hon'ble Governor of West Bengal opened the Programme at Tamluk in Midnapore district on 15th November2006.The State Animal Resources Development Ministers,Government of West Bengal and Kerala were also present.A seminar on the role of Panchayet and women participation in Animal Resources Development were organized.On the merit of performances of already developed indicators district prizes were distributed.Best Animal Friends were also awarded.

Table showing the ranking of different districts in Animal Resources Development.

Indicators	Rank-1st	Rank-2	Rank-3rd
Milk production	Nadia	Hooghly	Bardhaman
Egg production	24-Parganas(South)	Murshidabad	Birbhum
Artificial Insemination	Nadia	24-Parganas(North)	Siliguri(Darjeeling)
Animal Treatment	Jalpaiguri	Siliguri	Howrah
Green Fodder cultivation	24-Parganas(North)	Nadia	West Dinajpur
Over all success	Nadia		

The National Seminar held from 1st to 3rd February 2007 held in Kolkata also resolved for overall development of Animal resources in India and a model of Artificial Insemination(A.I) Workers at private level.

15 States of India and 150 delegates participated and inaugurated by Mr.Somnath Chatterjee,Hon'ble Speaker of Parliament,India.

1. Doorstep delivery of A.I with all inputs,services like animal health care,milk marketing facilities,supply of feeds and fodder,training of farmers etc. can only be ensured round the clock through out the year by trained private A.I. workers like Animal Friends.
2. Quality A.I services should be assured to the farmers' doorstep through providing quality A.I inputs and proper training to the Private A.I Workers.
3. Each State should formulate Cattle and Buffalo Breeding policy for development of cattle and buffalo wealth in respective State.Necessary legislative support for implementation of the policy may also be considered.Overall technical monitoring and vigil over maintainance of Breeding Policy in the State should be looked after by the Directorate.
4. Livestock resources mapping should be attempted and done as early as possible and conservation of native germplasm should be taken simultaneously by the state governments.
5. One A.I centre (including private A.I.worker) should be established for 800 breedable cattle /buffalo.There should be one Veterinarian for 6,000 breedable animals i.e 16000 large animals.This pattern is advocated for every State except for isolated areas in islands,hills,desserts etc.There should have uniformity in all the States in engagement of Veterinarian and Private A.I workers.
6. All the agencies engaged in doorstep delivery of A.I services in the State should procure FS Straws only from one window i.e the State Implementing Agency of the State concerned according to norms fixed under Breeding Policy of the State.Necessary legislative support for implementation of the policy may also be considered.

7. Attempts should be taken to provide to Private A.I.Workers not only out of the A.I done but also emphasis should be given to provide incentives on the basis of calf born followed by identification with ear tag and proper records.
8. There should be uniformity in establishment and maintenance cost of Private A.I Workers in all States.(This has a reference to circular no.Q-13021/3/2006-Agri dated 22-05-2006 of Planning Commission,Government of India).
9. Private A.I Workers may be considered as 'Facilitator'in Self-Help-Group (SHG or SGSY) formation and may be utilized as coordinator in co-operative formation.
10. Private A.I Workers are to be encouraged to form 'Growth Centre'for livestock and poultry development in his area,which in future may also act as Micro-Training-cum-Demonstration unit for the farmers.
11. Heifer rearing scheme and calf rearing scheme requires to be reintroduced by Govt.of India under NPCBB by providing funds to the States to boost the cattle and buffalo development programme.
12. Private A.I.Workers may be encouraged to act as a service providerd covering various activities such as Small Dairy units with fodder plots(to serve as demonstration unit),Agri Business Centre (supply of inputs like feed,fodder seeds),vermin-compost units,milk marketing,insurance,etc.
13. Private A.I.Workers may also be considered for various activities financially supported by Credit Linked Capital and /or interest Subsidy/Venture Capital Fund Scheme(NABARD).
14. Computerisation network,feed back system,data analysis and evaluation at regular interval should be ensured.The Private A.I.workers are the base level information provider.So special care,technical guidance should be assured to get correct relevant information in time from them.

15. For successful implementation of developmental programme,special care should be given to combat the reproductive diseases and to reduce the calf mortality.

MAN and Women Population(in million) engaged in different sectors of Rural India

- **Women** **Men**
- Agriculture1,870 75 ,470
- Dairying75,000 5,000
- Fisheries1,000 1,800
- Small-animal husbandry15,000 2,000
- Khadi and village industries1,700 1,990
- Handicrafts540 2,200
- Sericulture800 1,200
- Hand looms2,980 4.480
- Source: Various departments of the Government of India.

CONCLUSSION

India should have taken lesson how developed countries developed animal agriculture.

As in USA how animal agriculture developed; steps on the part of India might have a suitable lesson.

Are you a livestock producer? Do you know what you would do if you were notified that there was an outbreak of a foreign animal disease in the United States?

An outbreak of a disease such as foot and mouth disease would be devastating to the country's livestock industry, according to James Lane, Ford County Kansas Undersheriff.

"Terrorism is a low risk, high impact issue. We want to focus on the day to day criminal activity that is occurring and realize that this is about risk management," said Lane.

"There are so many factors to take into consideration when setting up a plan like this," Lane explained. "It includes involvement from livestock producers, veterinarians, the health department, trucking companies, packing plants, feedlots, the department of transportation, law enforcement, the media and others."

While many groups may have a plan, Lane stressed that it is important to know how to exercise it and know what to do.

Center Dedicated to Protecting Agriculture

The National Agricultural Biosecurity Center at Kansas State University was established in 2002 in response to Sept. 11. The center is dedicated to protecting the nation's agriculture infrastructure by planning and training producers and industry if a threat would occur.

NABC's initiatives include planning, training, outreach and research activities related to threat and risk analyses, incident response and detection and prevention technologies. The center coordinates agriculture biosecurity activities with federal, state and local agencies as well as other university and strategic partners.

"The exercises that have been done with different agencies are important, but the vital work is done before and after these tests," said Marty Vanier, associate director of the center. "Much planning happens before the exercise. After the exercise the groups involved can see what went right and wrong and what worked and didn't work."

Patrick Webb, National Pork Board veterinarian said that since every production system is different, producers should work with their herd veterinarian on reviewing or developing a current biosecurity and security plan.

Webb said that good biosecurity practices can help reduce the risk of disease introduction and spread within a production system. "It can also help contain a disease that is introduced and help to reduce the risk of further spread outside of a production system," he added.

He said that the National Animal Identification System(ID) would support animal health authorities and the industry in rapidly containing and eradicating an incursion of a foreign animal disease.

"The ID system would also help trace the problem back to the site of origin and would also help keep producers in the loop with timely communications on the situation and receive disease control measures necessary to protect their farms," Webb said.

Jon Caspers, vice president of the United States Meat Exports Federation Board and pork producer from Swaledale, Iowa said that his group knows it is pertinent to be able to track a situation back to its origin and feels that an animal ID program would make that happen much quicker.

"If the country affected by a disease such as FMD, it would be devastating to export markets, therefore the faster we find a problem the easier it is to control it if we know where it started," Caspers added. He also stressed the importance of good border patrols and proper trade and public policy regarding animal health issues.(Jennifer Bremer)

Similarly Animal friend in India will be facilitater and communicator in Animal resources development in India.

GFAR conference 2006 held in Delhi,the private sector delegates observed weakness of private sectors to represent the confererence though there is immense possibilities of growth for private investment in this sector.DFID also observed a strong scope of market in veterinary vaccines with increase in animal resources productivity.Resolved for private participation for Insurance and Financial Institutions.

The Chart showing Foods of Animal Agriculture in India

Milk	92 million tonnes
Meat	5.9 million tonnes
Eggs	44 billion
Poultry Meat	1.8 million tonnes

Veterinarians play an important role in monitoring animal resources'productivity.There are 7,872 Veteinary Hospitals/

Polyclinics,17,820 Veterinary dispensaries and 25,433 Veterinary aid centers/mobile dispensaries to look after animal health and assist in disease monitoring and surveillance.

There are about 250 disease diagnostic laboratories for quick and reliable diagnosis of animal diseases.About 26 Veterinary biological centers are for production of vaccine and diagnostics. There are 35 Veterinary Colleges and five Veterinary Universities in India to produce veterinary professionals.These are mostly financed by public sectors.

Private sectors take the services of veterinarians to grow by better participation both in micro(Small and Medium) and macro-businesses development causing overall socio-economic development.Animal welfare issues are important in the production of quality of meat and international Trade.There are laws,bills and orders for food safety and acts on prevention of Cruelty to Animals.

An important sector is ICT4D,Electronic medium likeTVchannels and Print media are growing rapidly by Public Private Participation in livestock sector development in India.

According toD-G Council of Scientific and Industrial Research, Bio-suite of Public sector developed in India was a result of a combined public-private enterprise.Private sector devotes less time and investment for doing original research and it is always outsourced from universities and research institutes.

Meat Trade in India

India has the largest livestocks comprising of 189 million cattle(About 80 million unproductive),96 million buffaloes,60 million sheep,120 million goats,185 million layers and 1,100 million broilers.

Total production of meat 5.9 million tones and 1.8 million tones of poultry meat ranking 5th position in the world.

The contribution of buffalo meat in India 25%(Table-I) which accounts for 1.2% of total world meat production(Table-II).

Table-I: Buffalo Meat Production 2003-2006 in 1000 Metric Tonnes

Year	Total Meat Production	Buffalo Meat Production	% of Share
2003-04	5898	1471	24.94
2004-05	5921	1582	
2005-06	5889	1592	

The demand for buffalo meat is high as it is relatively lean with fat content below 20% and has low cholesterol.It is cheapest meat in India because of availability at affordable prices as its demand is low as compared to other meats.

Table-II: Buffalo Meat Production in some of the Asian Countries(2003)

Country	Total Meat in1000MT	Buffalo Meat in 1000MT	Buffalo Meat as% of Total Meat	% of Animals slaughtered
India	5940	1466	24.7	11
Pakistan	1908	582	30.5	15
China	70898	305	0.4	17
Asia	102176	2846	2.8	13
World	253475	3154	1.2	13

Source of Table-I&II FAO(2004)

The lagest importer of Indian meat is Malayasia followed by the Philippines in the year 2005-06, 82.337 MT of Rs.4,994/-crores and 48,918MT of Rs.2,660/-crores respectively.

So,Private Public Participation in this sector will encourage coverage for food security and better Malnutrition Index by the policy makers utilizing veterinary scientists and huge other menpower exist in India.

Agriculture

WAYS BEEN INDIA'S most important economic sector. In the mid-1990s, it provides approximately one-third of the gross domestic product (GDP—see Glossary) and employs roughly two-thirds of the population. Since independence in 1947, the share of agriculture in the GDP has declined in comparison to the growth of the industrial and services sectors. However, agriculture still provides the bulk of wage goods required by the nonagricultural sector as well as numerous raw materials for industry. Moreover, the direct share of agricultural and allied sectors in total exports is around 18 percent. When the indirect share of agricultural products in total exports, such as cotton textiles and jute goods, is taken into account, the percentage is much higher.

Dependence on agricultural imports in the early 1960s convinced planners that India's growing population, as well as concerns about national independence, security, and political stability, required self-sufficiency in food production. This perception led to a program of agricultural improvement called the Green Revolution, to a public distribution system, and to price supports for farmers (see The Green Revolution, this ch.). In the 1980s, despite three years of meager rainfall and a drought in the middle of the decade, India managed to get along with very few food imports because of the growth in food-grain production and the development of a large buffer stock against potential agricultural shortfalls. By the early 1990s, India was self-sufficient in food-grain production. Agricultural production has kept pace with the food needs of the growing population as the result of increased yields in almost all crops, but especially in cereals. Food grains and pulses account for two-thirds of agricultural production in the mid-1990s. The growth in food-grain production is a result of concentrated efforts to increase all the Green Revolution inputs needed for higher yields: better seed, more fertilizer, improved irrigation, and education of farmers. Although increased irrigation has helped to lessen year-to-year fluctuations in farm production resulting from the vagaries of the monsoons, it has not eliminated those fluctuations.

Food-grain production increased from 50.8 million tons in fiscal year 1950 to 176.3 million tons in FY 1990. The compound growth rate from FY 1949 to FY 1987 was 2.7 percent per annum. Overall, wheat was the best performer, with production increasing more than eightfold in forty years. Wheat was followed by rice, which had a production increase of more than 350 percent. Coarse grains had a poorer rate of increase but still doubled in output during those years; production of pulses went up by less than 70 percent. The increase in oilseed production, however, was not enough to fill consumer demands, and India went from being an exporter of oilseeds in the 1950s to a major importer in the 1970s and the early 1980s. The agricultural sector attempted to increase oilseed production in the 1980s and early 1990s. These efforts were successful: oilseed production doubled and the need for imports was reduced. In the early 1990s, India was on the verge of self-sufficiency in oilseed production.After independence in 1947, the cropping pattern became more diversified, and cultivation of commercial crops received a new impetus in line with domestic demands and export requirements. Nontraditional crops, such as summer mung (a variety of lentil, part of the pulse family), soybeans, peanuts, and sunflowers, were gradually gaining importance.

The per capita availability of a number of food items increased significantly in the postindependence period despite a population increase from 361 million in 1951 to 846 million in 1991. Per capita availability of cereals went up from 334 grams per day in 1951 to 470 grams per day in 1990. Availability of edible oils increased significantly, from 3.2 kilograms per year per capita in FY 1960 to 5.4 kilograms in FY 1990. Similarly, the availability of sugar per capita increased from 4.7 to 12.5 kilograms per year during the same period. The one area in which availability decreased was pulses, which went from 60.7 grams per day to 39.4 grams per day. This shortfall presents a serious problem in a country where a large part of the population is vegetarian and pulses are the main source of protein.

There are large disparitics among India's states and territories in agricultural performance, only some of which can be attributed to differences in climate or initial endowments of infrastructure such as irrigation. Realizing the importance of agricultural production for economic development, the central government has played an active role in all aspects of agricultural development. Planning is centralized, and plan priorities, policies, and resource allocations are decided at the central level. Food and price policy also are decided by the central government. Thus, although agriculture is constitutionally the responsibility of the states rather than the central government, the latter plays a key role in formulating policy and providing financial resources for agriculture.

Land Use

In FY 1987, field crops were planted on about 45 percent of the total land mass of India. Of this cultivated land, almost 37 million hectares were double-cropped, making the gross sown area equivalent to almost 173 million hectares. About 15 million hectares were permanent pastureland or were planted in various tree crops and groves. Approximately 108 million hectares were either developed for nonagricultural uses, forested, or unsuited for agriculture because of topography. About 29.6 million hectares of the remaining land were classified as cultivable but fallow, and 15.6 million hectares were classified as cultivable wasteland.

These 45 million hectares constitute all the land left for expanding the sown area; for various reasons, however, much of it is unsuited for immediate cropping. Expansion in crop production, therefore, has to come almost entirely from increasing yields on lands already in some kind of agricultural use (see table 26; table 27, Appendix).

Topography, soils, rainfall, and the availability of water for irrigation have been major determinants of the crop and livestock patterns characteristic of the three major geographic regions of India—the Himalayas, the Indo-Gangetic Plain, and the Peninsula—and their agro-ecological subregions (see fig. 5; Principal Regions, ch. 2). Government policy as regards irrigation, the introduction of new crops, research and education, and incentives has had some impact on changing the traditional crop and livestock patterns in these subregions. The monsoons, however, play a critical role in determining whether the harvest will be bountiful, average, or poor in any given year. One of the objectives of government policy in the early 1990s was to find methods of reducing this dependence on the monsoons.

Himalayas

The Himalayan region, with some 520,000 square kilometers of land, ranks well behind the other two regions in agricultural importance. Despite generally adequate rainfall, the rugged topography allows less than 10 percent of the land to be used for agriculture. The sandy, loamy soils on the hillsides and the alluvial clays in the region's premier agricultural subregion, the Vale of Kashmir—located in the northwestern part of the state of Jammu and Kashmir—provide fertile land for agricultural use. The main crops are rice, corn, wheat, barley, millet, and potatoes. Most of India's temperate-zone fruits (apples, apricots, cherries, and peaches) and walnuts are grown in the vale. Sericulture and sheepherding also are being undertaken. In the eastern Himalayan subregion, the soils are moderately rich in organic matter and are acidic. Although much of the farming is done on terraced hillsides, there is a significant amount of shifting cultivation, which has resulted in deforestation and soil erosion. Rice, corn, millet, potatoes, and oilseeds were the main crops in the early 1990s. The region also is well known for the tea plantations of the mountainous Darjiling (Darjeeling) area in the northern tip of West Bengal.

Indo-Gangetic Plain

The vast Indo-Gangetic Plain, extending from Punjab to Assam, is the most intensively farmed zone of the country and one of the most intensively farmed in the world. Rainfall, most of which comes with the southwest monsoon, is generally adequate for summer-grown crops, but in some years vast areas are seared by drought. Fortunately, much of the land has access, or potential access, to irrigation waters from wells and rivers, ensuring crops even in years of drought and making possible a winter crop as well as a summer harvest. Wheat is the main crop in the west, rice in the east. Pulses, sorghum, oilseeds, and sugarcane are among other important crops. Mango orchards are common. Other fruits of the subregion include guavas, jackfruit, plums, lemons, oranges, and pomegranates. In the Great Indian Desert, rainfall is scanty and erratic. About 20 percent of the total area is under cultivation, mostly in Haryana and Gujarat states, and comparatively little in Rajasthan. The Indira Gandhi Canal—begun in 1958 as the Rajasthan Canal—was designed to bring water from the north. Progress was slow, and only the first stage was close to completion by the end of the Seventh Five-Year Plan (FY 1985-89). By then, the canal had substantially increased the area under cultivation in Rajasthan, and a new completion date of 1999 is anticipated (see Development Programs, this ch; Development Planning, ch. 6). The cultivable area is expected to expand further with the development of the canal's second stage during the 1990s. The leading crops of the subregion are millet, sorghum, wheat, and peanuts. Vast expanses of sparse vegetation provide sustenance for sheep and goats. In the late 1980s, dairy farming became important in locations that had sufficient pastureland.

Peninsular India

The east and west coasts, the coastal plains, and the deltaic tracts that extend inland for some 100 to 200 kilometers in Peninsular India benefit from both the June-to-September southwest monsoon and the October-to-November northeast monsoon. Farther inland, as the topography and climate change, so does the pattern of agriculture. The proportion of land under cultivation ranges from about 50 percent along the coastal plain and in the western part of Andhra Pradesh to about 25 percent in eastern Madhya Pradesh. Except in areas of certain developed river valleys, double-cropping is rare. Rice is the predominant crop in high-rainfall areas and sorghum in low-rainfall areas. Other crops of significance along the east coast and in the Central Highlands in the early 1990s were pigeon peas, mustard, peanuts, millet, linseed, castor beans, cotton, and tobacco.

On the Deccan Plateau, deep, alluvial black soils that retain moisture for a long time are the basis for much of the region's output of farm products. However, the region also has many farming areas that are covered by thin, light-textured soils that suffer quickly from drought. Whether a crop is made or lost is, therefore, often dependent on the availability of supplementary water from ponds and streams. About 60 percent of the land in the state of Maharashtra was under cultivation in the early 1990s, less in Madhya Pradesh. About 75 percent of the cropland of the Deccan during this period was planted in food crops, such as millet, sorghum, rice, wheat, and peanuts; most of the remaining cropland was planted in fodder crops.

In the far south of the Peninsula, the area under cultivation varies from about 10 percent in the Western Ghats, to 25 percent in the western coastal tract, to 55 percent on the Karnataka Plateau. Here is the India—the land of spices—that Vasco da Gama and other European navigators came searching for in the fifteenth century. On the Karnataka Plateau, sorghum, millet, pulses, cotton, and oilseeds are the main crops on the 90 percent of the cultivated land that is dry-farmed; rice, sugarcane, and vegetables predominate on the 10 percent that was irrigated in the late 1980s. Coconuts, areca, coffee, pepper, rubber, cashew nuts, tapioca, and cardamom are widely grown on plantations in the Nilgiri Hills and on the western slopes of the Western Ghats.

Land Tenure

Matters concerning the ownership, acquisition, distribution, and taxation of land are, by provision of the constitution, under the jurisdiction of the states (see Local Government, ch. 8). Because of the diverse attitudes and approaches that would result from such freedom if there were no general guidelines, the central government has at times laid down directives dealing with the main problems affecting the ownership and use of land. But it remains for the state governments to implement the central government guidelines. Such implementation has varied widely among the states.

Landholding Categories

India is a land of small farms, of peasants cultivating their ancestral lands mainly by family labor and, despite the spread of tractors in the 1980s, by pairs of bullocks. About 50 percent of all operational holdings in 1980 were less than one hectare in size. About 19 percent fell in the one-to-two hectare range, 16 percent in the two-to-four hectare range, and 11 percent in the four-to-ten hectare range. Only 4 percent of the working farms encompassed ten or more hectares.

Although farms are typically small throughout the country, the average size holding by state ranges from about 0.5 hectare in Kerala and 0.75 hectare in Tamil Nadu to three hectares in Maharashtra and five hectares in Rajasthan. Factors influencing this range include soils, topography, rainfall, rural population density, and thoroughness of land redistribution programs.

Many factors—historical, political, economic, and demographic—have affected the development of the prevailing land-tenure status. The operators of most agricultural holdings possess vested rights in the land they till, whether as full owners or as protected tenants. By the early 1990s, there were tenancy laws in all the states and union territories except Nagaland, Meghalaya, and Mizoram. The laws provide for states to confer ownership on tenants, who can buy the land they farm in return for fair payment; states also oversee provision of security of tenure and the establishing of fair rents. The implementation of these laws has varied among the states. West Bengal, Karnataka, and Kerala, for example, have achieved more success than other states. The land tenure situation is complicated, and it has varied widely from state to state. There is, however, much less variation in the mid-1990s than in the postindependence period.

Independent India inherited a structure of landholding that was characterized by heavy concentration of cultivable areas in the hands of relatively large absentee landowners (zamindars—see Glossary), the excessive fragmentation of small landholdings, an already growing class of landless agricultural workers, and the lack of any generalized system of documentary evidence of landownership or tenancy. Land was important as a status symbol; from one generation to the next, there was a tendency for an original family holding to be progressively subdivided, a situation that continued in the early 1990s. This phenomenon resulted in many landholdings that were too small to provide a livelihood for a family. Borrowing money against land was almost inevitable and frequently resulted in the loss of land to a local moneylender or large landowner, further widening the gap between large and small landholders. Moreover, inasmuch as landowners and moneylenders tended to belong to higher castes and petty owners and tenants to lower castes, land tenure had strong social as well as economic impact (see *Varna*, Caste, and Other Divisions; Settlement and Structure, ch. 5).

By the early 1970s, after extensive legislation, large absentee landowners had, for all practical purposes, been eliminated; their rights had been acquired by the state in exchange for compensation in cash and government bonds. More than 20 million former zamindar-system tenants had acquired occupancy rights to the land they tilled. Whereas previously the landlord collected rent from his tenants and passed on a portion of it as land revenue to the government, starting in the early 1970s, the state collected the rent directly from cultivators who, in effect, had become renters from the state. Most former tenants acquired the right to purchase the land they tilled, and payments to the state were spread out over ten to twenty years. Large landowners were divested not only of their cultivated land but also of ownership of forests, lakes, and barren lands. They were also stripped of various other economic rights, such as collection of taxes on sales of immovable property within their jurisdiction and collection of money for grazing privileges on uncultivated lands and use of river water. These rights also were taken over by state governments in return for compensation. By 1980 more than 6 million hectares of waste, fallow, and other categories of unused land had been vested in state governments and, in turn, distributed to landless agricultural workers.

Land Reform

A major concern in rural India is the huge number of landless or near-landless families, many of whom are wholly dependent on a few weeks of work at the peak planting and harvesting seasons. The number of landless rural families has grown steadily since independence, both in absolute terms and as a proportion of the population. In 1981 there were 195.1 million rural workers: 55.4 million were agricultural laborers who depended primarily on casual farm work for a livelihood. In the early 1990s, the rural work force had grown to 242 million, of whom 73.7 million were classified as agricultural laborers. Approximately 33 percent of the employed rural workers were classified as casual wage laborers.

Because of the large number of landless farmers and the frequent neglect of land by absentee landlords in the early years of independence, the principle that there should be a ceiling on the size of landholdings, depending on the crop planted and the quality of the land, was embodied in the First Five-Year Plan (FY 1951-55). An agricultural census was conducted to provide guidance in setting such ceilings. During the Second Five-Year Plan (FY 1956-60), most states legislated fixed ceilings, but there was little uniformity among the states; ceilings ranged from six to 132 hectares. Certain specialized branches of agriculture, such as horticulture, cattle breeding, and dairy farming, were usually exempted from ceilings.

All the states instituted programs to force landowners to sell their over-the-ceiling holdings to the government at fixed prices; the states, in turn, were to redistribute the land to the landless. But adamant resistance, high costs, sloppy record keeping, and poor administration in general combined to weaken and delay this aspect of land reform. The delays in legislation allowed large landowners to circumvent the intent of the laws by spurious partitioning, sales, gifts to family members, and other methods of evading ceilings. Many exemptions were granted so that there was little surplus land.

To ensure more uniformity in income, to combat evasion of the intent of the laws, and to secure more land for distribution to the landless, the central government in the 1970s pushed for greatly reduced ceilings. For a family of five, the central government guidelines called for not more than 10.9 hectares of good, irrigated land suitable for double-cropping, not more than 10.9 hectares of land suited for one crop annually, and not more than 21.9 hectares for orchards. Exemptions were continued for land used as cocoa, coffee, tea, and rubber plantations; land held by official banks and other government units; and land held by agricultural schools and research organizations. At the option of the states, land held by religious, educational, and charitable trusts also could be exempted. To protect the states from legal challenges to their land reform laws, the constitution was amended in 1974 to include in its Ninth Schedule the state laws that had been enacted in conformance with national guidelines. Land reform laws enacted after 1974 also were included in the amendment.

By the beginning of the 1990s, all states and union territories, except Goa, Arunachal Pradesh, Nagaland, Manipur, Mizoram, and Tripura, had passed ceiling laws to conform to central government guidelines. In Maharashtra, for example, the revised ceiling law that became effective in 1975 set upper limits at perennially irrigated land, 7.2 hectares; seasonally irrigated land, 10.8 hectares; paddy land in an assured rainfall area, 14.6 hectares; and other dry land, 21.9 hectares. By the early 1980s, about 150,000 hectares had been declared surplus under this act, about 100,000 of which had been distributed to 6,500 landless persons. A 1973 land reform amendment in Bihar set a range of ceilings on holdings for a family of five, from six to eighteen hectares depending on land quality, and offered an allowance for each additional family member, subject to a maximum of one-and-one-half times the holding. Within five years, the Bihar government had acquired 94,000 hectares of surplus land and had distributed 53,000 hectares to 138,000 landless families. Success nationwide was limited. Of the 2.9 million hectares of land declared surplus, nearly 1.9 million hectares had been distributed by the end of the seventh plan, leaving 1 million hectares still to be distributed as of early 1993.

By the early 1990s, nearly all the states had enacted legislation aimed at the consolidation of each tiller's landholdings into one contiguous plot. Implementation was patchy and sporadic, however. By the early 1980s, the work had been completed only in Punjab, Haryana, and western Uttar Pradesh and had begun in Orissa and Bihar. In most of the other states, nothing had been accomplished by the early 1990s. The Sixth Five-Year Plan (FY 1980-84) set a goal for the completion of the consolidation of holdings within ten years, which was not achieved.

In order to protect tenants from exorbitant rents (often up to 50 percent of their produce), the states passed legislation to regulate rents. The maximum rate was fixed at levels not exceeding 20 to 25 percent of the gross produce in all states except Andhra Pradesh, Haryana, and Punjab. The states also adopted various other measures for the protection of tenants, including moratoriums on evictions, minimum periods of tenure, and security of tenure subject to eviction on prescribed grounds only.

By the early 1980s, most of the cultivated area had been surveyed and records of rights prepared. In most states, revenue assessment—the tax on land—against farmland had been revised upward in keeping with a rise in farm prices. In several states, steps were taken to associate village assemblies, or *panchayat*, with the maintenance of land records, the collection of land revenue, and the management of lands belonging to government; the results of these efforts have frequently been unsatisfactory.

8

SCOPE FOR PROTO-TYPE EXPERIMENTS IN THE DIVERSE INDIAN AGRO-ECOSYSTEMS

CONCEPT

Precision agriculture, newly emerging agricultural management concept, embodies the convergence of biotechnologies & other agricultural technologies with space and informatics. With a goal to achieve the quantum jump in agricultural productivity, reduced cost of cultivation, diversified and resilient agricultural systems, the precision agriculture plays catalytic role in order to achieve a common ground between environmental and economic goals. It is basically designed to optimize agricultural inputs viz., fertilizers, pesticides, water etc, in tune with micro-level/field requirements. Optimization is focused on increased yields, reduced cost of cultivation and to minimized environmental impacts through location-specific management.

The success stories pertaining to Precision Agriculture have mainly drawn from the developed countries; wherein Agriculture is characterized by highly mechanized and automated systems, and is driven by market forces and has been professionally managed enterprise. Taking into account the predominance of fragmented land holdings, heterogeneity of crops and livestock and concept of farm families in the rural conditions, the model of Precision Agriculture representing the typical Indian Agricultural scenario is yet to evolve. While the ecological integrity of farming systems is an imperative need, it is equally important to extend the access of information and market to the small and marginalized farmers. The Precision Agriculture model for India while addressing these issues should provides an innovative route for sustainable agriculture in globalize and liberalized economy.

TECHNOLOGY ELEMENTS

Recent advances in technology for variable rate technology (VRT), with concurrent advances in remote sensing, GIS & GPS, and the developments taken place in crop simulation modeling, have provided enough opportunities and scope to take-up proto-type Precision Agriculture experiments. The VRT applies production inputs at rates appropriate to soil and crop conditions within micro-level field conditions. The VRT systems have been demonstrated for several materials, including herbicides, fertilizers, insecticides and seeds.

Role of space technologies becomes more crucial in order to address the spatial variability of soils and crops across the various scales of mapping. The space technology inputs also capture the vulnerability and dynamism of agricultural systems. The developments in space-borne imaging sensors, particularly their spatial, spectral and temporal resolutions are well characterized to capture these features. While high spatial resolution images enable mapping and monitoring the structural attributes of agro-ecosystems, high spectral resolution or hyper-spectral imaging addresses their functionalities. The high temporal resolution captures the dynamisms of agro-ecosystems.

The use of remote sensing, GIS and GPS for crop monitoring, condition assessment and yield modeling has already been well established. Crop simulation models (CSMs) provide potential production under the different scenario of constraints, including weather, soils, crops, cultural practices etc. The conjunctive use of VTR, remote sensing, GIS, GPS and CSMs provides technological framework for Precision Agriculture.

Components and Framework

Precision agriculture, basically, is characterized by reduced cost of cultivation (through optimization of inputs), improved control and increased resource use efficiency, through appropriate applications of Management Information System (MIS). While the reduced cost of cultivation is achieved through optimization of agricultural inputs taking into account economic push and environmental pull related factors, the control mechanisms are introduced by the help of VRT systems, model outputs and conjunctive use of remote sensing, GIS and GPS. The MIS comprises Decision Support Systems (DSS), collateral inputs and associated GIS databases on crops, soils & weather. Dynamic remote sensing inputs on in-season crop conditions, crop simulation model outputs on the potential production under the different constraining scenario, and the networks of labs and farms, form the essential ingredients of MIS. Increased efficiency does not employ only efficient resource use but also reflects in terms of less waste generation, improved gross margin and reduced environmental impact.

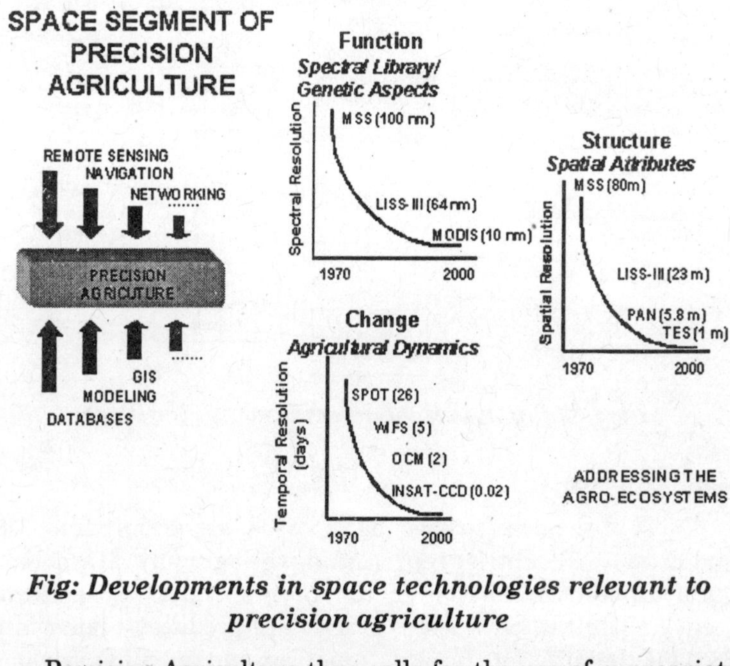

Fig: Developments in space technologies relevant to precision agriculture

Precision Agriculture thus calls for the use of appropriate tools and techniques, within a set of the framework as mentioned, to address the micro-level variations between crop requirements and applications of agricultural inputs. Inevitably, it integrates a significant amount of data from different sources; information and knowledge about the crops, soils, ecology and economy but higher levels of control require a more sophisticated systems approach. It is not simply the ability to apply treatments that are varied at the local level but the ability to precisely monitor and assess the agricultural systems at a local and farm level. This is essentially to have sufficient understanding of the processes involved to be able to apply the inputs in such a way as to be able to achieve a particular goal not necessarily maximum yield but to maximize financial advantage while operating within environmental constraints.

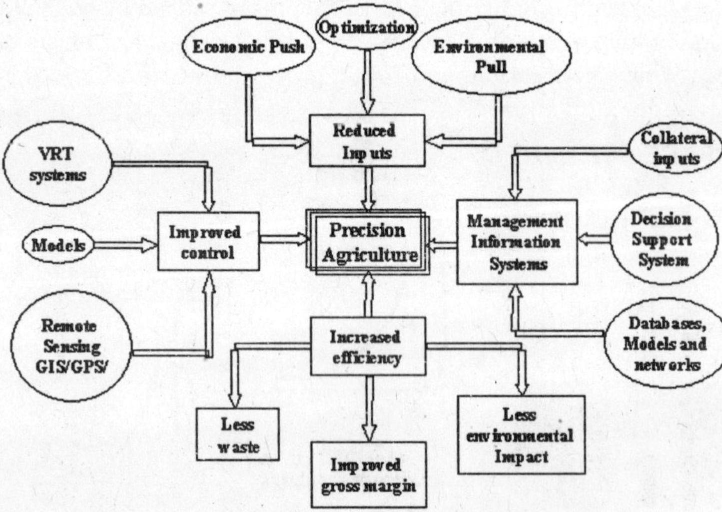

Fig: Components of Precision Agriculture

Case Studies

Most of the case studies, as reported, are mainly from US and European countries representing the agricultural systems, quite different from that of the India's typical agricultural scenario. Unless few success stories are produced taking into account the different farming practices and cropping systems available in India, Precision Agriculture practices cannot be replicated.

Models of Precision Agriculture drawn from US & UK A framework of the precision agriculture system, being followed in US & UK, is depicted in Figure above. At the core of the system is a GIS database, which are knowledge based and form the part of decision-making. The GIS databases include the following layers: field topography, soil types, surface drainage, sub-surface drainage, soil testing results, rainfall, irrigation, actual chemical application rates, and or even more frequently. The GIS enables a study of the relationship between these layers of information to determine cause and effect and to base decisions upon this knowledge.

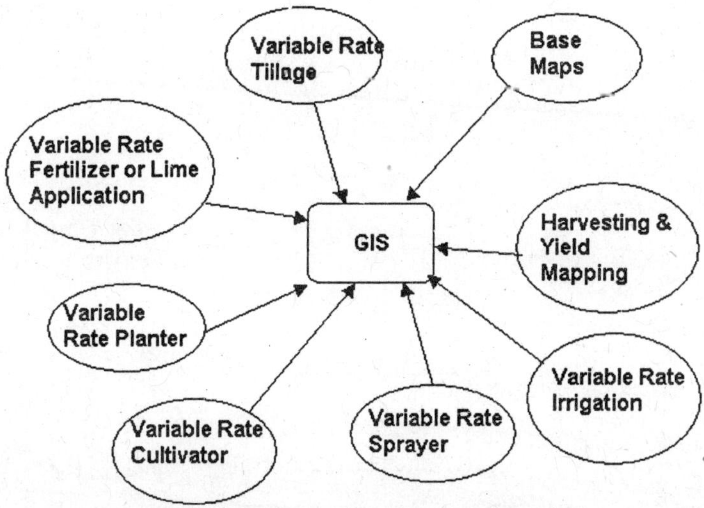

Fig. Precision Farming Overview

The main components, which make up a variable rate application system, are shown in Figure 2.2. Not all systems will necessarily contain all of the components shown. As variable rate technology develops, other system components may be included. The central component of variable rate application equipment is the computer/controller. This device receives information from several sources, which will in turn be used to control the application equipment. The controller may receive information from the application equipment and other sensors to maintain a database on the actual application rate as a function of field position Each field operations are governed by VRT systems. Tillage depth varies according to field location; for example, sub-soiling depth is dependent on field location. Seeding rates varies according to field location, which depends on factors such as topography and soil type.

Fertilizer application rates vary in relationship to factors such as soil type and the results from either real time or pre-application testing. Application of insecticides is dependent on insect location from either scouting reports or from aerial imaging. In like manner, the application of all inputs to the crop production process varies with field location.

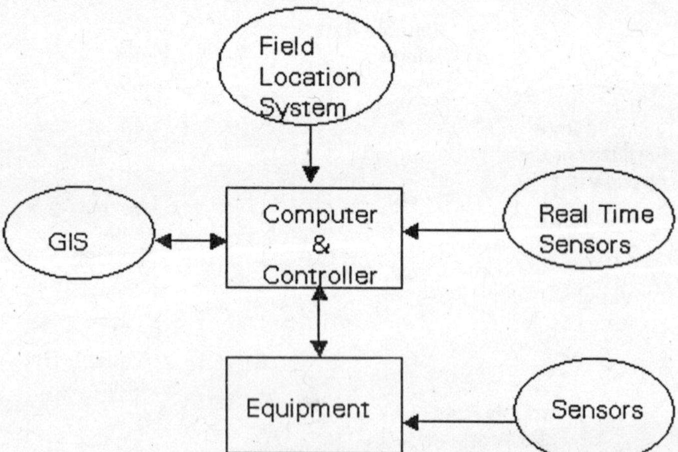

Fig: Components of VTR systems

There are two methodologies for implementing precision, or site-specific, farming. Each method has unique benefits and can even be used in a complementary, or combined, fashion:

The first method, Map-based, includes the following steps: grid sampling a field, performing laboratory analyses of the soil samples, generating a site-specific map of the properties and finally using this map to control a variable-rate applicator. During the sampling and application steps, a positioning system, usually DGPS (Differential Global Positioning System) is used to identify the current location in the field.

The second method, Sensor-based, utilizes real-time sensors and feedback control to measure the desired properties on-the-go, usually soil properties or crop characteristics, and immediately use this signal to control the variable-rate applicator

Remote Sensing technologies are used for in-season crop condition assessment including the crop moisture or nutrient stress and other conditions—indicating the need for irrigation and fertilizers or insecticides. All of these data give farmers more opportunities to tailor their management decisions to their farm's needs. These inputs help the farmers to locate and analyze the stressed part of the field with reduced sampling in map-based technique.

9

Space Applications to Agriculture in India

Agriculture has been at the top of our priorities for space applications. The road map of agricultural applications started with the first remote sensing experiment on identification of coconut root-wilt diseases in Kerala using Infrared aerial photography- way back in 1969-70. Since then, space applications to agriculture sector have touched almost all the segments of agricultural ecosystems. These include the mapping and monitoring of major crops, soils/degraded lands, command areas, wastelands, surface & ground water, floods & drought, and watersheds.

Agricultural statistics that provides the vital informatics base to agriculture sector originates from age-old village level Patwari system. It often moves upward with its inherent subjectivity and bias. There are hundreds of crucial decisions at different levels, which are taken purely based on this. In order to strengthen the foundations of agricultural statistics in the country, a remote sensing based the nationwide mission called pre-harvest Crop Acreage and Production Estimation (CAPE) was launched in late 80s. Covering all the major cereals, pulses and oilseeds, CAPE provides in-season crop statistics with 90/90 accuracy at state level. The CAPE could thus provide the scientific basis of agricultural statistics and is transitioning further to yet another institutional destination entitled Forecasting Agricultural Output using Space Agro-meteorology and Land-based Observations (FASAL) within Ministry of Agriculture itself.

Synthesizing the state-of-the-art in econometrics, agricultural meteorology and remote sensing based modeling, FASAL envisages the multiple productions forecasting of the major crops with improved accuracy to the extent of 95/95 criteria, timeliness and scope in terms of covering the whole country. The FASAL could be used to provide scientific solutions facilitating crop insurance, bridging the gaps between crop production and post-harvest technology, pricing and policy decisions.

Extensions of irrigation, genetically improved crops and use of inorganic fertilizers have been the harbingers of Green Revolution in the country. Those multi-purpose irrigation commands, which accelerated the irrigation networks in 70s & 80s, are now characterized by depleted irrigation efficiency, water logging and salinity. At the behest of Ministry of Water Resources, remote sensing based command areas inventory has been taken up in the priority irrigation commands to enhance the water use efficiency. In the similar line, a coarse scale land degradation mapping was carried out at behest of Ministry of Agriculture. It is important to mention here that in some of critically affected areas of Gangetic plains of Uttar Pradesh, land degradation to the extent of village level has been mapped out. Based on this, several land reclamation efforts- including the World Bank supported programmes, have been taken up to restore the health of soils. In coming years, with the support from Ministry of Agriculture, ISRO/DOS is planning to launch Nationwide Land Degradation Mapping Mission at the scale 1: 50,000, so as to delineate land degradation at Village level.

By virtue of the unique combinations of IRS satellites with varied resolutions and capabilities, the space segment in the country is in a position to monitor agricultural drought and recurrent floods more efficiently. Capturing the flood events with the damages to the Villages and crops has formed the scientific basis for damage assessment and thereby relief operations at various levels. This year, for example, flood inundation maps in Orissa were generated depicting the marooned areas during the different flood waves and disseminated to the user community within hours through the

Internet. Space inputs to agricultural drought monitoring is yet another aspect to target the districts and taluks on the basis of severity of drought.

Department of Space at the behest of Ministry of Rural Development has carried out nation-wide wasteland mapping on 1: 50,000 scale using the IRS data. The Wasteland Atlas of India covering the wasteland statistics of entire country has been brought out. These maps would help in retrieving the information at village/watershed (500 ha) levels, for implementation of wastelands/watershed development programmes. The uniqueness of wasteland mapping has also been the creation of digital database with village as well as watershed boundaries.

With the hydro-geomorphological maps prepared under the National Drinking Water Mission, and subsequently to the Rajiv Gandhi National Drinking Water Mission, satellite remote sensing has really made a dent in reaching its potentials down the line to grassroots. Search for groundwater, particularly in areas with consolidated and semi-consolidated rock formations, considered more difficult from the point of view of exploration as well as recharging of groundwater, is considerably aided by the use of the hydro-geomorphological maps. These maps are extensively used for locating prospective groundwater sites around problem habitations in the country, as a part of the 'scientific-source finding'.

At the behest of the Planning Commission, scientific inputs were generated from remote sensing on land use/land cover of entire country to evolve agro-climatic zonation based planning. These applications have moved further to large scale mapping across the problem regions of the country (especially the rain-fed regions), towards micro-level decision making for sustainable development. Under remote sensing based Integrated Mission for Sustainable Development Mission (IMSD), land and water resources development plans were carried out for about 80 mha. in 175 districts of the country. Implementation of these plans in certain areas has enriched the ground water potential, increased the cropping intensity along with the net returns from the fields. It is important to note that these experiments were conducted in the terrain of low productivity, mainly dry lands/rain-fed areas, predominantly wastelands, where the scope of doubling food production is the main issue. Emphasis is being laid in operational utilization of this vast database for ground level implementation by concerned user agencies.

The success of these experiments lies not only due to the applications of space inputs alone, but largely due to the synergistic interfaces among grass root level farmers, administrators/policy makers and multi-disciplinary scientific communities. With operationalisation of the National Natural Resources Management System (NNRMS) in the country, with DOS as the nodal agency, the space, ground and user segments have been properly tuned to respond to thc challenges of sustainable crop production in the country.

The INSAT-VHRR observations on clouds, cyclone depressions, surface radiance and monsoon parameters in the Indian Ocean have really strengthened the agro-meteorological services in the country. To disseminate the appropriate locale-specific agricultural packages, there are several programmes such as 'Krishi Darshan' being beamed through all the regional channels. An operational system of public instruction has been established since 1995, using the INSAT based Training and Development Communication Channel (TDCC) for disseminating the improved agricultural practices, training primary school teachers, Panchayat Raj, elected representatives, Anganwadi workers, wasteland development functionaries etc. The Jhabua Development Communication Project (JDCP), undertaken by ISRO in Jhabua district of MP, has been providing communication support to the developmental activities and also interactive training to the development functionaries, besides empowering the poor and marginalised tribal.

A pilot project sponsored by the Planning Commission, a single window information service provider concept - called Agro-climatic Planning and Information Bank (APIB) has been developed for the farmers of drought prone areas of Karnataka. The APIB is an effort to empower the farmers and reduce their vulnerability to crop failures. Moving to the farmer's cooperatives, ISRO has taken up jointly with IFFCO - a major initiative on developing remote sensing and GIS based decision support system for fertilizer movements in the priority areas. Finally, space applications over last three decades in agriculture have made this sector information rich and have addressed the various issues encompassing the different sectors of agriculture.

PRECISION AGRICULTURE: IS REALLY NEEDED IN INDIA'S CONTEXT?

Agriculture in India, as we see today, is at the crossroads.

1. On the one hand there are depletions of ecological foundations of the agro-ecosystems, as reflected in terms of increasing land degradation, depletion of water resources and rising trends of floods, drought and crop pests and diseases. On the other hand, there is imperative socio-economic need to have enhanced productivity per units of land, water and time.
2. At present, 3 ha of rain fed areas produce cereal grain equivalent to that produced in 1 ha. of irrigated. Out of 142 million ha. Net sown areas, 92 million ha. are under rain-fed agriculture in the county.
3. From equity point of view, even the record agricultural production of more than 200 Mt is unable to address food security issue. A close to 60 Mt food grains in the storehouses of Food Corporation of India (FCI) is beyond the affordability and access to the poor and marginalized in many pockets of the country.
4. Globally, there are challenges arising from the Globalization especially the impact of WTO regime on small and marginalized farmers.
5. Some other unforeseen challenges could be anticipated global warming scenario and its possible impact on diverse agro-ecosystems in terms of alterations in traditional crop belts, micro-level perturbations in hydrologic cycle and more uncertain crop-weather interactions etc.

At this stage, agriculture needs new paradigms to deal with the present situation. The strategy lies in integration of the dynamic information and scientific knowledge into the management of agro-ecosystems, and thereby optimizing the radiation, water and nutrient usages. Agriculture has to transition from high inputs material inputs to the optimum level, through the appropriate use of information, knowledge and strategies for efficient resource usages. In such case, productivity of agricultural may not be the function of the quantum of agricultural input use alone, but will include information, knowledge and efficiency while managing the agricultural practices.

With the fragmentations of land holdings and predominance of small and marginal farmers, our agricultural systems are basically dis-aggregated farm families. Fundamentally, Precision Agriculture aims at a dis-aggregated micro-level farm management strategy with intense information inputs addressing the variations of soils, crops, water, chemicals, market access etc. Taking into the present state of Agriculture in the country, Precision Agriculture is absolutely essential in order to address poverty alleviation and food security to a very large cross-section of the population. For example, at present, the average Orissa farmer produces slightly more than one ton of rice per hectare, and keeps little below one tone for his family. But if he can produce four tons, then he has three tons to sell, and more cash in hand. The smaller the farm, the greater the need for a marketable surplus.

Defining Precision Agriculture in Indian Context Taking into account the typical characteristics of Indian Agriculture, the definition of Precision Agriculture, therefore, must encompass the strategy and framework to achieve higher productivity, reduced cost of cultivation by optimization of inputs, and diversified & resilient agricultural systems. All these goals are to be achieved within the typical constraints of India's agro-ecosystems. While the depletion of ecological foundations of farming systems needs to be arrested, the access of information, credit, agricultural inputs and market to the small and marginal farmers are equally crucial. The Precision

Agriculture model for must be derive encompassing all these issues, within a broad framework of addressing the negatives of globalization and achieving sustainable agriculture in the long run.

The definition of Precision Agriculture has to be micro-level contextual and should capture the local variabilities, vulnerabilities and dynamisms of agro-ecosystems. For example, Agriculture in Punjab, Haryana & Western UP is characterized by higher productivity (about 4 t/ha), higher use of inputs (irrigation ~ 96%; fertilizer consumption 0.158 t/ha), higher cropping intensities and predominance of medium and large farmers. Sustainability of su h agro-ecosystems has been a cause of concern in the recent times. The model for Precision Agriculture for such ecosystems may focus on sustainability through the optimization of agricultural inputs and thereby reduction of cost of cultivation, and cropping system analysis.

Agriculture in Southern part of the country-especially Andhra and Tamilnadu is moderate yield (around 2 t/ha), inputs (irrigation ~ 55%; fertilizer consumption 0.12 t/ha) and cropping system based, with predominance of small and marginal farmers. Precision Agriculture for such system should aim at enhancing the productivity based on in situ soil & water conservation. The Eastern India especially Orissa, though high potential agricultural system is unfortunately characterized by low inputs (irrigation ~ 25%, fertilizer use 0.025 t/ha), low yield (~1.2t/ha) and low cropping intensity, with predominance of marginal and small farmers. A model for Precision Agriculture for Orissa could focus more on poverty alleviation and food security through enhanced productivity and low cost of cultivation. The models for Precision Agriculture in Indian context could therefore be the micro-level and contextual - addressing the empirical issues in the diverse agro-ecosystems of the country. The access and outreach of best practices in Precision Agriculture - as demonstrated by contextual models, to the large cross-sections of the farming community would then help in cultivating the new paradigms in age-old agriculture enterprise in the country.

Precision Agriculture models are not complete, unless the parameters related to empowerment of the farmers; especially small and marginal farmers are integrated. In this context, ISRO has also initiated Gramsat project in Orissa. In the line of JDCP, the Gramsat project aims at empowering the people especially the poor and marginalized, by awareness building and access to information and services. Towards this, a network of one-way video and two-way audio Village Information Kiosks is being developed in the selected blocks of Orissa. The same networks are also planned to facilitate e-governance in the region. Precision Agriculture model should present a synergy that could lead to a holistic mission, focused on agricultural development with the backdrop of present issues and challenges.

DEVELOPMENT IN AGRICULTURE AND BIOTECHNOLOGIES

Sustainable Survival

All living organisms are driven by two motivating forces: to reproduce themselves and to survive. Survival depends largely on protection from their enemies and having access to sufficient food. Consequently, for countless millennia the developmental history of hominids has to do with their search for sustenance and survival: an uneven progression from hunting and gathering to crop cultivation and livestock husbandry, the discovery and devising of biotechnologies by which to increase, protect, preserve and transform natural and cultivated biological species and materials into food, fibre and other goods of personal, domestic or commercial utility.

History records that agriculture has progressed as much through empirical perception as from scientific intention. In today's world, intensive large-scale highly mechanised production systems exist together with simple labour intensive subsistence systems where a few crops and farm animals survive on less than a half-hectare of land.

Some nations and regions, which for many centuries were prosperous, maintained efficient profitable agricultural systems and produced well in excess of their needs, subsequently lost or forsook their agricultural capabilities and descended into a state of food insufficiency.

To be sustainable, agriculture requires a complex of critical resources, not least human skills and experience, and the dedicated political will of governments and those who legislate each nation's priorities. Libya and Nigeria are two nations where the discovery of oil has led to gross neglect of agriculture. In other ecologies, changes in weather patterns or excessive exploitation of natural resources and depredation of ecosystems have together degraded once fertile lands into near deserts.

Agriculture and Biotechnologies

During the first century BCE, a Roman farmer called Varro is reported to have said: "Agriculture is a science which teaches us what crops should be planted in each kind of soil, and what operations are necessary in order that the land will produce the highest yields in perpetuity". The fourth century Bishop Ambrose of Milan wrote in *De Officiis* that agriculture is the only way of making money that gives no offence.

'Agriculture' is derived from the Greek word 'áãñïæ' which means 'arable land', and 'áãñïíìôá' (English equivalent: 'agronomy'): the management of land. 'Agriculture' relates literally to the cultivation and management of land, but in popular usage embraces virtually all patterns and scales of crop and livestock production, including horticulture, (literally: the cultivation of gardens: Latin '*hortus*' = garden), more generally used to designate the cultivation of fruits and vegetables. 'Agroforestry' or 'Agri-sylviculture' relate to the integration of crop and livestock production with natural and cultivated arboreal species, in some instances the production and utilisation of non-timber forest products such as tree fruits, nuts, wild honey and medicinal plants. 'Agribusiness' has come to embrace most industries that provide goods, machinery and services to agriculture, and the industries that preserve, process, store, distribute and market the products from agriculture.

Agriculture, horticulture and biotechnologies are predominantly devoted to human health: the production of food to provide energy for healthy activity, bodily growth and restoration, drugs to diagnose, prevent and cure diseases, and fibres from which garments and fabrics are created.

'Biotechnology' is derived from several Greek words: 'âôïæ' meaning 'life',: 'an industrial art or craft', and 'systematic application of an industrial art or craft'. In recent literature, 'biotechnology' (singular) is frequently used exclusively to connote genetic modification of organisms and related modern biochemical reactions. In reality 'biotechnologies' (plural) embrace all processes that serve to protect, preserve, extract and convert materials of biological origin into products of industrial, commercial, social, economic or hygienic value and utility. Milling, baking and fermentation processes, devised by ancient Egyptians and Babylonians are as correctly designated 'biotechnologies' as are synthetic and extractive processes of biological materials that derive therapeutics and diagnostics from genetically modified organisms. One could cite instances of 'biotechnology' used to describe processes that have not progressed beyond a laboratory or field experiment, which more accurately should be designated 'biosciences'. In the subsequent text the plural noun 'biotechnologies' denotes the broader concept defined above.

Evolution of Planet Earth

The history of Planet Earth's existence can be classified geologically, according to the changing structure and composition of the earth's outer crust, or by progress in the development of tools, implements and technologies. Geological timing starts with the Precambrian period (about 4,500 million years ago), succeeded by the Cambrian (about 600 million years ago), which is believed to coincide with the appearance of the earliest fossils. Of particular interest to humans and their immediate ancestors is the Pleistocene period that began about 2 million BCE, during which period the Earth's climate underwent fluctuations between warm and cold temperatures, cycles that have continued until the present Holocene period that began around 15,000 BCE. The Holocene began with the end of the final glaciation, believed to have lasted from about 60,000 to 15,000 BCE. One of our many hominid ancestors, Neanderthal Man, a cave dweller and hunter, who appeared about 200,000 BCE, survived by hunting and gathering during the Pleistocene period.

From an agricultural perspective, an Earth-time classification based on the tools hominids used is perhaps more meaningful. It begins some 2.5 million years ago with the Early Palaeolithic period when hominids used primitive sharp-edged stone tools to harvest vegetation, cut up carcasses and sharpen sticks. During the Mesolithic period that began about 15,000 BCE, stone tools came with multiple sharp edges, attached to handles fashioned out of wood or deer antlers. The Mesolithic era merged into the Neolithic when more refined and efficient tools were made of polished stone, used as axe-heads to cut down trees, as sickles in crescent shapes with handles to harvest crops, with sharp blades to scrape animal hides. Harvested grains were ground by stone pestles and mortars, the pestle a heavy stone club, the mortar probably a concave hollow in a rock.

Hunting and Gathering

Several scholarly narratives describe how our ancestors progressed from hunting and gathering to settled crop cultivation and animal husbandry. Blaxter writes that to plan for the future we must build on knowledge accumulated from long past experience. The immediate past can provide a basis for simple extrapolation and adaptation of recent trends, while human history over the distant past can illustrate many of the difficulties that have confronted, and destructive practices that have constrained, humans in their quest for an adequate food supply. Blaxter discusses how changing weather patterns, massive epidemics, natural and man-made instabilities have been encountered and thus, in a broad sense, indicate the potential difficulties that lie ahead. Critchfield illustrates how settled agriculture, the practice of irrigation, domestication of animals and the invention of the plough led to the early establishment of village communities and their eventual expansion into towns and cities.

All animals in the wild survive by hunting and gathering; as, more than one million years ago, did the earliest hominids: the Australopithecines (African 'Ape-men'), the Pithecanthropoids distributed over Africa, Europe and Asia between

500,000 and 1,000,000 years ago. *Homo habilis* and the Neanderthals (*Homo neanderthalensis*) hunted and gathered from *ca* 200,000 to 35,000 BCE and devised Mousterian flint tools, more advanced and efficient than those of their predecessors. The first modern humans: *Homo sapiens* probably appeared between 10,000 and 15,000 years ago, being preceded by countless earlier primitive hominids.

Anthropological opinions differ on when the various hominid species first appeared on earth and when some became extinct. *Homo erectus*, so named because of an apparent ability to walk upright, may have existed more than 1.5 million years ago. It is speculated that the Neanderthal hominids (*Homo neanderthalensis*) arrived about 200,000 years ago. Scientists at the University of Utah have studied evolution of certain head lice in tracing the history of early hominids. They conclude that *Homo neanderthalensis* and early *Homo sapiens* lived contemporaneously until about 25,000 years ago, when the Neanderthals and some other early hominids disappeared. The Utah scientists conclude that *Homo sapiens*, possessing larger brains and possibly bigger bodies may have destroyed their more primitive hominid cohabitants.

Hunting and gathering persists as a source of sustenance for many primitive people. Survival by hunting and gathering is sustainable if the dependent humans do not increase in numbers faster than the natural rate of food source regeneration. It is speculated that as each hunting-gathering tribe exhausted its immediate natural food resources it would migrate to find other sources of supply. Whenever a population grew faster than the natural regeneration of the wild animals and plants that provided its food some or all would die of starvation. It is not improbable, as has been the case in recent times, that the least productive, young children and old men and women, would be the first to be sacrificed in times of shortage.

Early Hominid Diets

The ways in which prehistoric hominids acquired their food and the composition of their various diets can only be conjectured, accurate determinations being virtually impossible.

To begin, sites for suitable study must be found. In dry environments biological remains are better preserved, so more sites have been studied in arid and semi-arid ecologies than in the humid tropics. Samples suitable for analysis are plant and animal parts that have resisted decay for several thousand years. Fleshy, non-lignified plant parts rarely survive for any length of time. Carbonised seeds and highly lignified seed coats are among the best survivors. Bones of very small animals and of fish are less resistant to degradation than large animal bones and mollusc shells. Consequently, what can be analysed illustrates only an unknown proportion of the foods consumed by ancient hominids.

Palaeoclimatology, carbon dating, pollen analyses, isotopic and elemental analyses of hominid skeletal remains are methods variously employed to discover some of the components of ancient hominid diets. The ratio of ^{13}C to ^{12}C in collagen from hominid bones indicates plant and animal sources that were dominant in the hominid's diet [Price 1989]. The ratio of ^{15}N to ^{14}N helps to distinguish between leguminous and non-leguminous diets. Where written or pictorial records are available, as from the early riverine civilisations of Egypt, Mesopotamia and the Indus valley, dates and patterns of the transition from hunting-gathering to settled farming can be more reliably determined. Nonetheless, present analytical methods cannot distinguish between plant and animal remains from a wild or a domesticated species. Evidence of lined storage pits and stone pestles and mortars do not irrefutably indicate that the grains stored and processed were cultivated, since both could be used for grains harvested from the wild. Out of the many animals slaughtered by hunters, only a few species appear to have been domesticated. Studies and samples analysed from the Mesolithic and early Neolithic periods show that wild red deer and roe deer were abundant, though neither seem to have been domesticated.

Middle Palaeolithic hominids, whose remains are spread across Europe, West Asia, the Middle East and parts of Africa, were more successful hunters than their predecessors. Animal remains suggest they devoured many species of large mammals;

how much of their meat came from hunting and killing live animals, how much from the scavenging of dead carcasses is unknown. Sites in Europe where bones from several animal species were found may have been primitive abattoirs where hominids processed the spoils of their kill. Archaeologists have discovered sharpened pointed stones probably used as spear heads, indicating a new kind of hunting technology. Simple hearths at many Middle Palaeolithic sites show that fire was used to cook meats, to dry and cook grains and, with protective clothing from animal hides, enabled Neanderthals to spread to and survive in such colder regions as the Central Asian steppes.

The Mesolithic period extended from the end of the Pleistocene ice age until the Neolithic, about 10,000 years ago, when settled farming first appeared in different locations and at different times. The end of the Pleistocene ice age led to rapid environmental change where warmer post-glacial conditions of the Holocene epoch caused ice sheets to retreat and sea levels to rise. Temperate forests spread across Europe and Asia, large herds of mammals such as reindeer were replaced by red deer, roe deer and wild pig. Reindeer, elk and bison, adapted to cold climates, retreated to the north; the mammoth, giant deer and woolly rhinoceros became extinct.

From Hunting to Settled Agriculture

Settled agriculture must be regarded as among the most important innovations and developments in human history. For whatever reasons particular communities changed from hunting-gathering to agricultural technologies devised and developed from perceptive empiricism, each adopted system being influenced by local soil, climate and resource conditions. There seems general agreement among archaeologists and ethno-botanists that transitions from hunting and gathering to settled farming occurred in different ways, according to different patterns and stimuli in different places. Settled agriculture, in which food-crop seeds were planted and animals domesticated began during the Neolithic period around 10,000 years ago. In some communities settled farming seemed to result from local indigenous development, in others farming technologies devised or discovered in one location radiated outwards to be adopted or adapted by other communities.

To evolve from gathering to crop cultivation required the discovery of indehiscent plant species. In order to reproduce themselves, wild grasses and other plants when mature scatter their seeds, some of which will then take root. Seed scattering plants are described as dehiscent (= bursting or splitting open). This trait of dehiscence is particularly noticeable in sesame, hence the expression 'Open sesame'. Settled cultivators have need of indehiscent plants so that, when mature, the seeds can be harvested and stored, to be eaten or later planted.

The earliest farmers were of Neolithic culture located in the region now covered by Iraq, Iran, Jordan, Syria and Turkey, along the Nile in Egypt, the Indus river valley and its tributaries in what is now India and Pakistan. Exactly when the first crops were derived from planted seed is uncertain and varied among regions. Though most are believed to predate 6000 BCE, some ethno-botanists believe the earliest may date from around 10,000 BCE. According to carbon isotope dating, wheat, barley and possibly flax were cultivated in Mesopotamia from 8,000 BCE. Wheat and barley cultivation slowly spread northwest to reach Britain about 2,000 BCE. A diversity of fruits and vegetables: onions, melons, cucumbers, dates, figs were cultivated in Mesopotamia and nearby regions between the fourth and third millennia. During the second millennium, cotton was grown and spun in India and Egypt, linen and silk in China.

Examination of animal and plant remains by 14C indicates that sheep, bovines, goats, pigs and asses were domesticated in Mesopotamia between 9,000 and 6,000 BCE. Neolithic settlements were more permanent than the camps of nomadic hunters and gatherers though some early farming settlements had to move periodically if weather patterns changed or soils became infertile from excessive cultivation.

The first crop cultivation systems probably began by clearing land for continuous cropping until soil fertility was exhausted, at which time the cultivators would move to another location, which then and now in certain poor nations is the basis of slash-

and-burn agriculture. It is generally assumed that the early cultivators started by empirical discovery of which wild plants were edible and from which seed could be removed and planted on cleared land.

From the harvest of their cultivated crop, farmers learned to save some seed for the next planting. Repeated cultivation of a chosen genotype would eventually provide the farmer with a stable crop with relatively homogeneous characteristics. Herds of goats and sheep were assembled, captured young wild animals with the most useful characteristics, those with small horns and high milk yields, being domesticated. Their progeny would become progressively tame, with comparatively uniform physiologies.

Demographic, Economic and Social Change

Settled farming, with domestication of plants and animals led to profound social changes, particularly in regions such as Mesopotamia, Egypt and the Indian Harappan culture, where fertile soil, ample water for crop irrigation and abundant sunlight provided conditions for highly productive farming systems. Archaeologists have located early farming sites in Thailand and China, where rice and other indigenous crops were cultivated, in a few African territories where indigenous sorghum and millets were grown. Most comprehensively described were farm settlements along the Euphrates, Tigris, Nile and Indus valleys [Critchfield 1981; Thapar 1996].

The capacity to produce food surplus to the producers' needs, facilitated trade in agricultural commodities, the establishment and growth of non-farming urban communities. It also led to income stratification as prosperous farmers and urban traders increased their wealth. Cultivators in Mesopotamia and Egypt were probably the first to use systematic irrigation, to establish cities, towns and regulated social organisations. The Indian civilisations were not far behind. The development of cuneiform script in Mesopotamia and Egyptian hieroglyphics [the writing of priests] provided means of communication and records for posterity [Leick 2002; de Burgh 1947].

Mesopotamia

'Mesopotamia' means 'between rivers', the rivers being the Tigris and Euphrates that rise in the Anatolian mountains and flow south to the Persian Gulf. In contemporary geographical usage, Mesopotamia includes Iraq, eastern Syria, parts of Iran and much of south-eastern Turkey. It comprised two distinct ecological zones: the northern Fertile Crescent, where rain-fed and irrigated agriculture started and the first urban settlements appeared some 10,000 years ago; and the southern flood plain, the rich fertile land where, in dry seasons, crops were irrigated from an ingenious system of dykes and canals. The most enduring legacies of ancient Mesopotamia are the dozens of established cities, and the invention of writing. Each of the cities strung along the rivers controlled its own rural territories, its agricultural production and system of irrigation, its individual form of governance and administration.

Many of the ten principal cities – starting with Eridu about 8,000 years ago and on to Babylon, which around 2500 BCE grew from a small town to a magnificent city that lasted into the first centuries of the contemporary era – disappeared over time variously because of climatic change, rivers changing their courses, military conflict and reasons not fully understood. Most now exist only as archaeological remains [Leick 2002].

According to some historians, most of Mesopotamia was under water prior to the Chalcolithic period. This is possibly the origin of the Biblical flood mythology. There is an ancient Sumerian myth that the storm god Enlil decided to reduce the population by drowning them in a flood. The god of wisdom, Enki, persuaded a wise man Atrahasis (also known as Ziusudra) to build a boat and to stock it with seeds and species of all living organisms, so as to ensure the survival of all creatures when the flood subsided [Armstrong, K 2005].

As the climate gradually changed, some 6,000 years ago, the floods receded, and the plains became habitable. The post-deluvian virgin alluvial soil supported barley, wheat, flax, horticultural crops and livestock production. The southern marshlands were a substantial source of fish, game and wild life. With water from the Tigris, Euphrates and their tributaries, crops were irrigated during seasons of low rainfall. Thus, starting in the fourth millennium, 4000 to 3000 BCE, southern Mesopotamia [Babylonia, Akkad and Sumer] provided an extensive exploitable ecosystem able to produce abundant food, thus encouraging settlements which expanded into large urban areas. Produce from farmers, fisher-folk and hunters stimulated a transition from subsistence agriculture to rural-urban complexes. Donkeys and oxen were probably the main beasts of burden.

Urban Commerce

The urban traders, artisans and administrators were fed by abundant food from the surrounding cultivated land. Artisans developed pottery from native clays, textiles out of flax and wool of domesticated sheep. Administrators prescribed and enforced rules that restrained conflict, enabling people spread among the settlements to live together in comparative peace. Surplus grain, leather from sheep hides, sun-dried fish, dates, other fruits, textiles and timber from natural forests, and locally mined metals and semi-precious stones were exported to urban settlements along the navigable Tigris and Euphrates [Leick 2002].

Theophrastus, Aristotle's celebrated pupil, in his *History of Plants*, describes how in ancient Babylon wheat fields were harvested twice each year, beasts being subsequently fed on the residual stubble. He stated that on cultivated, irrigated land each planted wheat seed could produce between a 50 and 100-fold harvest. One assumes he was comparing the total weight of biomass harvested not the weight of harvested seed to the weight of the seed planted. On fertile soil irrigated from nearby rivers, Babylonian farmers could harvest two crops every year, sufficient to satisfy their own needs and those of the expanding urban populations. [de Burgh 1947].

Egypt and the Nile

The Greek historian Herodotus wrote: "Egypt is the gods' gift of the Nile". Through Egypt, the Nile flows some 1,200 km from the first cataract in the north to the broad delta where it empties into the Mediterranean.

The Egyptian Nile is fed from the White Nile, which originates in northern Uganda, supplemented by the Blue Nile, which flows in at Khartoum, and the Atbara which joins about 450 km north of Khartoum.

Both the Blue Nile and the Atbara originate in the Ethiopian highlands. Over many centuries the abundance of the Nile, the fertile silt deposited after each annual flood, and Egyptians skilful management of dykes and irrigation systems supported a flourishing agrarian society. Diverse plant species were cultivated for food, clothing and shelter, well into the period when, under Roman rule, Egypt was regarded as the breadbasket of the Mediterranean [de Burgh 1947; Livy 1960; Cowell 1948]. The Nile sustained Egypt's fishing industry for several thousand years.

The Egyptian population grew from several hundred thousand during the pre-dynastic period (5000–3000 BCE) to several million by 1000 BCE. Egypt's cities and settlements were located close to the Nile, the densest population being along the flood plains. During most years when the rains in Uganda and the Ethiopian highlands were abundant, the Nile overflowed its banks. The floodwaters as they receded deposited a rich, black fertile soil to sustain productive crop cultivations over many millennia [McEvedy 1976].

History records the decline and fall of many once prosperous and advanced nations. Egyptians were literate, numerate and practised productive irrigated agriculture more than 5,000 years ago. They devised a decimal numerical system and the means to communicate in writing. The mathematics of dynastic Egypt was significantly influenced by agriculture. Calculations of area, length × breadth, were used to determine the size of farmers' fields, and before the fourth Millennium Egyptians calculated the value of ð [pi] the ratio of circumference to diameter as equal to 3.145. The present accepted value is 3.143. Knowing the value of ð enabled Egyptians to calculate the storage capacity of cylindrical grain bins.

Indus and Harappan Culture

The earliest traces of hominid activity in India go back to the second interglacial period between 400,000 and 200,000 BCE and show evidence of stone tools. There followed a long period of slow evolution that culminated in the spectacular Indus valley civilisation, known as the Harappan culture, which matured around 2500 BCE. The antecedents of the Harappan culture were village sites in the Baluchistan hills and rural communities along rivers that flowed in Rajasthan and Punjab. The Vedic civilisation, which began in the Indus valley, is said to have believed more in monuments of the mind and thought than in material monuments. Archaeological excavations have brought to light some 80 settlement sites dating from about 3500 BCE. The Indus civilisation was among the world's earliest; Indian historians wisely state that civilisation began when cereals and cattle were first domesticated. The Indus civilisation culminated in two main centres: Mohenjodaro close to the Indus valley, and Harappa on the Ravi river. Wheat cultivated near to Mohenjodaro is claimed to be an ancestor of the wheat varieties now grown in the Punjab. Both Mohenjodaro and Harappa maintained large granaries in which the annual wheat harvest was stored. In India cotton was first grown. Grain, cotton and other commodities were transported in wheeled carts drawn by oxen, and by river boats [Mookerji 1961].

The Harappan culture was essentially urban, the two main cities Mohenjodaro and Harappa being maintained from surplus crops produced by the surrounding rural farmers and stored in elaborate granaries. Harappan communities developed a flourishing trade with people in the Persian Gulf and Mesopotamia [Thapar 1966]. Some historians believe the first Sumerians to settle in Mesopotamia migrated from Baluchistan [Bouquet 1962].

Roman Agriculture

Marcus Tullius Cicero, Roman Consul, politician and philosopher, wrote: "Of all gainful occupations nothing is more pleasing and better becomes a well-bred man than agriculture" [Cowell 1948]. Some 6,000 years ago active volcanoes in Central

Italy covered the Latin plain with ash and lava that provided Roman farmers with an extremely fertile soil. For several centuries Roman farmers enjoyed exceptional prosperity. Between 265 and 70 BCE the number of male Roman citizens rose from 290,000 to 900,000. When women, slaves and non-citizens were counted, during the last century BCE, the total Roman population probably exceeded 3 million [Tacitus 1965]. But soil fertility and productivity could not keep pace with population growth, a constraint aggravated by top-soil stripped of its protective tree and grass cover being washed away by torrential rains. Animal manure and wood ash were insufficient to replace lost soil nutrients and what was once fertile crop land degenerated into land fit only for grazing animals.

To pursue Rome's military adventures and conquests, many of the best farmers were recruited into the Roman legions. By the turn of the millennium, as BCE changed to CE, Rome imported about 190,000 tonnes of wheat from Sicily, Sardinia and North Africa. Soon thereafter Egypt, the principal granary of the Mediterranean, became the main supplier of food grains to Republican Rome [Cowell 1956; Livy 1960].

Of the three civilisations whose early prosperity was derived from productive agriculture, only India remains as an efficient producer of food grains. As recorded in *Indian Agriculture 2003*, published by the Indian Economic Data Research Centre, since the 1950s India's annual wheat harvest has risen from 6 million to 70 million tonnes.

From Manual to Mechanised Processes

Virtually all the technologies employed to provide for basic human needs – agriculture, food processing, drugs, pottery and textiles – were discovered and developed empirically. Domestic and artisanal technologies were practised for many centuries before there was any scientific understanding of why they worked. Scientific understanding began in the late eighteenth and nineteenth centuries; only during the last century have novel biotechnologies been devised and developed based on scientific principles. For many centuries most apparent progress was from gradual, episodic mechanisation: the replacement of human hands by machines. Long ago iron replaced wood in simple ploughs, the primitive iron plough, stronger than wood, turned top-soil over rather than just scratching the surface.

Kates [1994] illustrates how new tools and technologies have stimulated human population growth over the past million years by raising available food supplies. Progressive improvement in hand tools used by hunters and gatherers resulted in an estimated population increase from 150,000 to 5 million during the period from one million to 100,000 years ago; a second surge from 5 million to *ca* 1 billion came with expansion of settled agriculture between *ca* 8000 BCE and the middle of the nineteenth century, since when the application of science to food production, preservation, processing and distribution has stimulated a world population explosion to more than 6 billion [Kates 1994; Hulse 1995].

Most rapid and far-reaching farm mechanisation spread across the Western Great Plains of the United States during the nineteenth century. The steel blade ploughshare appeared in 1830, the reaper in 1834, soon followed by mechanised harrows, seeders, threshers and binders, many manufactured and marketed by Cyrus McCormick. Mechanisation disruptively reduced the need for human labour on American farms: one McCormick threshing machine operated by four men replaced 400 farm labourers. Towards the end of the nineteenth century, guano from Peru and nitrates from Chile began to replace or supplement animal manure as nitrogen fertilisers. Introduction of the steam tractor accelerated farm mechanisation. During the second half of the nineteenth century more than 160 million hectares of new land came under cultivation in the United States [Nye and Morpurgo 1955].

As in ancient Egypt and Mesopotamia, rising farm productivity stimulated the growth of cities, the rate of American city expansion being unprecedented in human history. In 1910 Chicago's population reached 20 times the size in 1860. Mechanisation, more abundant farm productivity and rural-to-urban migration occurred in Europe and North America. During the twentieth century, the proportion of the Canadian work force employed on farms fell from *ca* 40 per cent to barely 3 per cent. The human misery suffered in cities that grew too rapidly, without planning or provision of essential services, is forgotten or ignored among nations of Africa, Asia and Latin America where urban expansion is uncontrolled, where slums and unhygienic environments are all too familiar.

From Empiricism to Scientific Agriculture

One of India's most distinguished scientists, the late Professor Y. Nayudamma, made the profound observation that science and technologies serve no useful purpose if they fail to provide for the needs of human beings. India's agricultural scientists and farmers in productive States have served its peoples exceptionally well, transforming the nation from dependence on imported grains and food aid to an abundant self-sufficiency in food grains.

Contrary to long-standing belief, agriculture does not inevitably progress starting with scientific research that begets invention, that leads to technological innovation, and ultimately results in ever-increasing crop production. Farming systems research (FSR), that starts by assessing and understanding farmers' adopted technologies, opportunities, constraints, resources and risks, began during the early 1970s. The benefits to farmers realised by FSR have been less widely publicised than research that generated higher yielding cereal crop phenotypes. It is through modifications based on careful systematic analysis and understanding of existing farming practices that sustainable systems are most effectively and sustainably developed, adapted and adopted. [IDRC 1981; Zandstra *et al* 1981].

Blaxter contends that modern food systems, highly dependent on energy other than solar radiations, are inherently unsustainable. As has come about over the past 12,000 years, novel or improved technologies and systems are needed to maintain an equitable, sustainable balance between food supply and persistently rising demand [Blaxter 1992]. Blaxter estimates that roughly 98 per cent of today's world's food supply is derived from terrestrial ecosystems. Though some 80,000 identified plants and countless animal species are believed to be edible, roughly 90 per cent of all human food calories come from only 15 plants and eight animal species.

Early Scientific Evolution

In his imaginative novel *Brave New World* Aldous Huxley asserted that the world's history of economic, technological and industrial development can be divided into two periods: a. before and b. after Henry Ford, the Detroit automobile industrialist who devised the continuous on-line process by which to assemble automobiles. In this author's opinion a more rational division is pre-and post-Faraday. If the Uruk-Sumerian invention of cuneiform script in the fourth Millennium, and Johannes Gutenberg's creation of the printing press in the 1450s can be considered the most important contributions to international scholarship, it seems appropriate that Michael Faraday started his working life as a bookbinder.

Though he never graduated from a university, Faraday's fascination with, and self-acquired knowledge of science led to his appointment as scientific assistant to Sir Humphrey Davy. Faraday's renowned contribution to electromagnetism began when he reviewed the Dane Hans Christian Oersted's discovery of the magnetic effect associated with an electric current. Faraday was the first person to discover benzene whose six-carbon ring structure, later elucidated by Kekule, is of fundamental importance to organic chemistry, to agricultural and food biochemistry [Schumm 1987].

Science Together with Empiricism

Although during the fourth century BCE Aristotle classified more than 500 animal species within hierarchies, and his pupil Theophrastus wrote a *History of Plants*, agricultural practices changed little from the time men and women began to cultivate crops and husband livestock around 10,000 years ago. Thales, one of the first Greek philosophers in the sixth century BCE, postulated that water was the main constituent of all plants. But it was not until the seventeenth century CE when van Helmont observed and reported the dependence on water of his experimental willow, Stephen Hales demonstrated how water is transmitted through roots and leaves, Joseph Priestley and Jan Ingenhosz described leaf respiration, and von Liebig and Dumas experimented with plant uptake of nitrogen and other nutrients, that an understanding of what influences plant growth began to emerge [Mees 1947; Gribbin 2002].

From the time of Faraday and his contemporaries in the nineteenth century scientific explanations of natural phenomena, of technologies empirically discovered, devised and developed over many millennia, gradually emerged. Nonetheless, while scientific underpinnings were coming to light, discovery and progress from thoughtful observation was widely evident.

During the eighteenth century an English farmer, Coke of Okeham, improved soil fertility with little help from chemists; while Robert Bakewell increased meat and wool per animal by selection and breeding of Leicestershire sheep more than 100 years before Mendel discovered the laws of heredity. The famous Rothamstead experimental station blossomed out of field experiments started by a private farmer John Lawes, who bore the total cost for almost 60 years until in 1889 a sustaining Trust Fund was endowed to carry on some of the world's longest systematic studies of soil fertility.

During the seventeenth century, Grew, Camerarius and Sprengel described plant male and female organs – stamens, styles, ovaries and ovules – and how insects carried pollen between plants.

Microscopy and Cytology

Who first discovered the magnifying power of convex lenses is unknown. The thirteenth century English scholar Roger Bacon is said to have constructed eye spectacles. During the seventeenth century Galileo devised a compound microscope, equipped with convex lenses he had ground. In 1625 he published microscopic details of various insects. In the late seventeenth century, a self-educated Dutch microscopist Anton van Leeuwenhoek, who earned his living in a drapery store, delicately ground glass lenses used to study and describe previously invisible microscopic materials. He described tiny parasites that infested fleas, a discovery that inspired Jonathan Swift to write:

So naturalists observe, a flea

Has smaller fleas that on him prey;

These have smaller ones to bite 'em

And so proceed *ad infinitum*.

Van Leeuwenhoek described other cellular organisms that he named 'animalcules', which probably were protozoa or bacteria. During the second half of the nineteenth century, bacteria were more accurately identified by Louis Pasteur, who demonstrated how micro-organisms could be thermally inactivated.

From Alchemy to Chemistry

Robert Boyle, son of the Earl of Cork, is most renowned for demonstrating the inverse relation between gaseous volume and pressure. Whether or not his book, *The Sceptical Chymist: Chymic-physical doubts and paradoxes*, provided the bridge between alchemy and chemistry, Boyle undoubtedly pioneered scientific method, inspired by Francis Bacon's thesis that scientific investigations must begin by critical assessment of all relevant available data, not by imagining a wonderful idea then seeking facts to sustain it. From the time of Faraday, progress in scientific method, in analytical chemistry, microscopy and cytology progressed to a greater understanding of living organisms, development of agricultural, food and pharmaceutical technologies.

The quarter-century following Faraday's empirical discoveries of electro-magnetic induction and the anaesthetising properties of ether was a vintage period in scientific progress. Darwin published *The Origin of Species*, Mendel formulated his laws of genetic inheritance, Maxwell published his theory of electromagnetism, Pasteur and Koch described the microbial origins of many diseases, Kekule elucidated the cyclical structure of benzene, Gibbs published his first laws of thermodynamics related to chemical reactions, Geissler's vacuum-tube technology paved the way for discovery of electrons ['cathode rays'], X-rays and the eventual discovery of radioactivity. A congress at Karlsruhe, convened by Kekule, led to a rational nomenclature for chemical compounds. The classification of elements according to their atomic weights by Cannizzaro, Newlands, de Chancourtoix and Meyer culminated in Mendeleyev's periodic table of all known elements. The later classification by atomic number did not greatly alter the Medeleyev table.

Joseph Black, Henry Cavendish, Joseph Priestley and Carl Scheele advanced the development of chemistry as a distinct science. Their discoveries of carbon dioxide, oxygen, hydrogen, ammonia, hydrogen chloride and sulphur dioxide; of the composition of water; and their finding that air is a mixture of gases, are illustrative of what was achieved by men of immensely diverse interests, most of whom, by contemporary scientific standards, would be classed as non-professionals [Gribbin 2002]. An attempted dissertation of all that science and technology have contributed to modern agriculture and biotechnologies lies far beyond the scope of this publication. Nevertheless, a few exceptional persons deserve special mention.

Lavoisier

Antoine-Laurent Lavoisier was born in 1743 in Paris, the son of a well-to-do lawyer. Lavoisier qualified as a Licenciate of Law, at the same time studying astronomy, botany, chemistry, geology and mathematics. He never practised law, his first work being a project to complete a geological map of France. During the late eighteenth century, Lavoisier devoted his talents to chemistry, demonstrating that when sulphur burns it gains weight, the first step in understanding how atmospheric oxygen (Priestley's 'pure air') is critical to combustion processes. From later experiments he concluded that animals derive their body heat from combustion processes dependent on inspired oxygen. He determined the weight of ice melted during ten hours by the heat from a guinea pig's body. Then he measured the weight of carbon burned to melt the same weight of ice. He measured how much 'fixed air' or CO_2 the guinea pig exhaled during ten hours and how much 'fixed air' was generated by burning different weights of charcoal. Lavoisier thus anticipated von Liebig's demonstration, 100 years later, that animals derive their body heat and energy from *in vivo* conversion of carbohydrates and lipids.

Lavoisier was the first to demonstrate that water is formed by the combination of hydrogen ('inflammable air') with oxygen, and how 'fixed air' is a compound of carbon and oxygen. His books *Traité elementaire de chimie* and *Méthode de nomenclature chimique*, published in 1787, laid the foundation for chemistry as a distinct scientific discipline, and provided the basis for a logical and systematic nomenclature for chemical elements and compounds: 'oxygen' replacing the earlier 'phlogisticated air', 'inflammable gas' giving way to 'hydrogen', 'oil of vitriol' being renamed sulphuric acid. One-third of the latter book was a dictionary of new chemical names presented in alphabetical order together with the old names, many coined by alchemists on the basis of the colour and physical appearance of the substance [Gribbin 2002; Crosland 1978; Ward and Dubos 1972].

Lavoisier's many profound pronouncements include: "La vie est une fonction chimique", demonstrably true in that all life processes are sustained by the conversion of one form of energy into another by biochemistry; and "Scientists should be as precise and meticulous in their written and spoken words as in their experimental measurements", advice all too often ignored in speeches and writings about 'sustainable development'. Due to an earlier involvement with Tax Farmers, in 1794 the French revolutionaries decreed that Lavoisier be executed by the guillotine.

Chemical Nomenclature

In 1806 Berzelius defined Organic Chemistry as: "That which describes the composition of living bodies and the chemical processes within them", the first distinction being between substances of vegetable and animal origin. Benefiting from Lavoisier's earlier publications, in 1860 Kekulé convened a congress in Karlsruhe at which a group of distinguished scientists discussed systems of nomenclature and symbols for organic compounds. From this congress and subsequent discussions Organic Chemistry progressed from Berzelius' definition to the present concept: the complex chemistry of thousands of carbon compounds [Crosland 1978; Dixon 1973]. It was not until the twentieth century that Berzelius' concept of organic chemistry evolved into the discrete discipline 'biochemistry', at first designated 'physiological chemistry'. From a typical dictionary definition: "Biochemistry is the study of the chemical constituents of living organisms, their functions, reactions and transformations" [Rose 1970].

Taxonomic Classification

Ray and Linnaeus

Arguably, the most productive naturalist of the seventeenth century was an Englishman, John Ray. Ray travelled extensively throughout Europe observing, describing and collecting thousands of specimens of natural flora and fauna. The results of his studies are reported in his books: *Ornithology*, *History of fishes* and *History of plants*. The latter, published in three volumes between 1686 and 1704, describes more than 18,000 plants classified according to distribution, natural habitat and those believed to possess pharmacological properties. Ray established 'species' as the basic units of taxonomic classification. His book "History of insects" was published posthumously.

Ray's taxonomy was based on anatomy, morphology, physiology and systematic principles adopted by the better known Carolus Linnaeus, born in 1707 in Sweden [Gribbin 2002]. The Linnaeus family name, Ingemarsson, was legally changed by his father to Linnaeus, the Latinised version of the Linden tree. Though he graduated in medicine, Linnaeus' life-long interest was in biology. He noted differences between the reproductive organs of flowering plants as a criterion of classification. He lived until 1778, switching from the Chair of Medicine to the Chair of Botany at Uppsala University in 1742. His most famous book, *Systema natural,* was first published in 1740. His binomial system of naming species was spelled out in Volume 1 of the tenth edition in 1758. Linnaeus classified close to 8,000 plant species and some 4,400 animals, all arranged in a hierarchy from 'Kingdom' at the top to 'species' at the bottom. As taxonomic techniques were refined, some of Linnaeus' original Latin binomials have changed; but his principles remain as the standard of species identification: a Latinised generic plus a specific name: e.g., *Canis lupus*, the scientific name for 'wolf'.

Linnaeus was the first to include humans in a zoological classification system. He questioned why humans are the sole species '*sapiens*' in the genus '*Homo*'. In his 1746 publication *Fauna Svecica* he writes: "I have yet to find any characteristics by which man can be distinguished on scientific principles from an ape." Based on modern DNA evidence man should be classified among the chimpanzees, possibly as '*Pan sapiens*'. The accepted classification '*Homo sapiens*' is used so as not to offend certain religious sensitivities [Ward and Dubos 1972; Gribbin 2002]. As modified by later taxonomists Linnaeus classification of animals now reads:

Kingdom	*Animalia*
Phylum	*Chordata*
Subphylum	*Vertebrata*
Class	*Mammalia*
Order	*Primates*
Family	*Hominidae*
Genus	*Homo*
Species	*Sapiens*

After Linnaeus died, James Smith, a wealthy English botanist, was instrumental in 1788 in founding the Linnean Society which, in London, holds most of Linnaeus' collections of biological specimens [Gribbin 2002].

Genetic Inheritance

The years in the nineteenth century when Faraday lived (he died in 1867) yielded an abundance of discoveries that advanced the development of agricultural science. In 1837,

Schleiden proposed that all plant tissues are composed of cells. A year later, Schwann suggested that all living organisms are of cellular composition. In 1858 Virchow, a professor of pathology in Berlin, asserted that all cells are derived by division of pre-existent cells. Using more powerful microscopes, Hertwig and others later discovered cell nuclei and, working with sea-urchins, observed how sperm penetrated into eggs, the nuclei of the sperm and egg cells fusing into a new nucleus that combined material from both parent cells.

Flemming and van Beneden reported the presence of thread-like structures in cell nuclei, which they named chromosomes because of their susceptibility to staining by chemical dyes. At Freiburg, Weismann described how when a cell divides its chromosomes are duplicated to be shared between the two daughter cells. Weismann declared chromosomes to be carriers of hereditary characteristics between parents and progeny (Hulse 2004).

Mendel

Now recognised as the founder of classical plant breeding, Mendel was born in 1822 into a farming family in Moravia (then in the Austrian empire). Being too poor to afford a university education, Mendel joined the priesthood where he changed his Christian name from Johann to Gregor. After completing his theological studies, recognising Mendel's exceptional intellect and intelligence, the abbot of his monastery in Brünn (now Brno in the Czech Republic) enabled Mendel to study at the University of Vienna. There he acquired a knowledge of physics, chemistry, statistics, probability mathematics and plant physiology.

After two years, and without a degree, Mendel returned to the monastery where, in 1856, he began a ten-year study of heredity in peas. Working with several thousand plants, on a plot 35 × 7 metres, Mendel transferred pollen by hand from one plant to another, meticulously recording his every action and the results. His results from hundreds of repeated experiments, all subjected to rigorous statistical methods learned in Vienna, revealed that when peas with rough wrinkled seed coats were crossed with smooth-coated seeds, 75 per cent of the progeny had smooth coats, 25 per cent were rough, wrinkled seeds. This phenotypic phenomenon of unequal inheritance of parental characteristics illustrated the now familiar principle of dominant and recessive genes.

Contrary to Darwin's theory of heredity, Mendel demonstrated that inheritance results not in a uniform blending of parental characteristics but from specific traits being inherited in unequal proportions.

Mendel's unique results were overlooked for more than half a century, in large part because in 1868 his abbot died and Mendel was appointed as the abbot's successor. His religious and administrative responsibilities allowed time neither to continue his plant breeding experiments nor to promulgate and disseminate his results to many who could have benefited. The rediscovery early in the twentieth century of Mendelian laws of inheritance, together with the elucidation of chromosomal behaviour, laid the foundations for highly productive plant breeding, most notably the so-called Green Revolution during the latter half of the twentieth century. [Stern and Sherwood 1966; Stoskopf 1993; Wortman and Cummings 1978; Gribbin 2002].

Closing Comment

The foregoing illustrates how biotechnologies were discovered, devised and developed over many millennia; how the basic biotechnologies – agriculture, food, drugs and textiles – began as domestic or artisanal crafts and gradually, over many millennia, were modified by perceptive empiricism. Early, primitive technologies preceded the most rudimentary scientific understanding by many hundreds of years, systematic scientific discovery being of very recent origin with much yet to be learned. What is known of the development of biotechnologies essential to human health and survival: agriculture, food, drugs and textiles, can truly be described as 'a gradual unfolding' and illustrate that sustainable progress can rarely be hastened but more reliably results from persistent patient systematic imaginative activity.

As is emphasised later, all development – to be rational, productive and sustainable – requires a systematic understanding and logical integration of all that influences and impinges upon the intended development. This is especially true of development related to the biotechnologies, upon which human life, biodiversity and their environments depend, biotechnologies which are progressing and changing more rapidly than at any time in the past.

ASIAN FARMERS

We do not know if cereal agriculture developed independently in India and Pakistan. Farmers were living at Mehrgarh west of the Indus River by 6000 B.C.E. Cereal crops, humped cattle, pig, and water buffalo were domesticated by local south Asian populations in this region. The new economies spread rapidly in northwest Pakistan and into other areas. By 5000 B.C.E., the Mehrgarh people were cultivating locally domesticated cotton, which became a vital trade commodity in later centuries, one of the foundations of later urban civilization in the Indus River Valley.

There are signs of intensive exploitation of wild plant foods in the Southeast Asian highlands, and perhaps even domestication of wild yams and other root plants as early as 8000 B.C.E., but the evidence is very controversial. The rices and Asian millets ancestral to modern rice were first domesticated somewhere between northeast India, Southeast Asia, and southern China. The initial process of domestication probably took place in an alluvial swamp area where there was plenty of seasonal flooding to stimulate crop growth. The earliest records of cultivated rice come from China's Middle Yangzi River Valley, dating from as early as 7000 B.C.E., but it is likely that similar early dates will come from the Ganges Plain and other regions in the future.

By 6500 B.C.E., Southeast Asians were moving from the hills onto river plains and into lowland areas, where intensive cultivation and irrigation permitted rice agriculture. Rice soon became the vital staple of farmers throughout southern and Southeast Asia, but its spread is not well documented. At

Homutu in coastal southern China, a community of rice farmers lived in a marshy area between 5000 and 4000 B.C.E. Their village of beautifully made wooden houses was surrounded by forests and was close to a great diversity of food resources. The women of Homutu made a distinctive type of cord-decorated pottery, which was also widespread in Southeast Asia, Taiwan, and Japan.

North Chinese farmers relied not on rice, but on local cereals and seeded plants such as millets, sorghum, and the mulberry planted on river valley soils. Again, agriculture was a local development. The earliest farming villages date to about 5000 B.C.E., perhaps earlier. For the next 2,000 years, the Yangshao farming culture flourished in the Yellow River Basin, an area as large as the early centers of agriculture in Egypt or Mesopotamia. Each Yangshao community was self-contained, overlooking a fertile river valley. By 3000 B.C.E., Yangshao people were enjoying a characteristic and thoroughly Chinese culture with its own naturalistic art style. The roots of Chinese cuisine and language may date from Yangshao times.

During these 2,000 years, many distinctive farming cultures developed throughout China. They became the Longshanoid cultures, which were the founder societies of early Chinese civilization.

THE ORIGINS OF FOOD PRODUCTION IN THE AMERICAS (C. 5000 B.C.E. AND LATER)

The Native Americans were remarkable for their expertise with native plants. They domesticated not only staples like maize and beans, but also hundreds of varieties of potato, amaranth, and other crops that are now international staples. In contrast to the Old World, only a few animals were available for domestication, among them the guinea pig, llama, and turkey. Potential beasts of burden had become extinct at the end of the Ice Age.

As in the Old World, experimentation with native plants began early, especially in the Andean area of South America, where people strove constantly to expand the range of wild plants into marginal environments. In North America, hunter-gatherer societies in major Midwestern river valleys were planting such native plants as goosefoot and marsh elder to supplement wild stands as early as 2500 B.C.E. And the Guila Naquitz excavations in central Mexico show that hunter-gatherers in Central America also experimented with bean and squash cultivation as a means of surviving dry years in an unpredictable environment as early as 7000 B.C.E.

Maize *(Zea mays)* was the staple cereal crop for many Native American societies when Christopher Columbus landed in the New World in 1492. It was cultivated from Argentina and Chile northward to Canada, from sea level to high in the Andes, in low-lying swamp environments, and in arid lands. Hundreds of races of domesticated maize evolved over the millennia, each with a special adaptation to local conditions. The wild ancestor of maize is thought to be teosinte, a wild grass that grows over much of Central America.

Teosinte was transformed into a primitive maize through human selection that was much easier to harvest. The earliest known maize cobs date to about 2750 B.C.E. in the Tehuacán Valley in southern Mexico and from the Valley of Mexico in the highlands.

By 1700 B.C.E., the inhabitants of the Tehuacán Valley were growing amaranth, gourds, and maize to the point that agriculture was the dominant part of the subsistence economy. Tehuacán is by no means unique, for many groups throughout Central America were turning to agriculture by this time. In time, a primitive form of domesticated maize with kernels in eight rows was the ancestral crop, which spread thousands of miles from its original homeland after 2700 B.C.E.

In South America, people living in the Andean highlands of Peru were cultivating potatoes, maize, beans, and squash by 2500 B.C.E., some of these crops, especially potatoes, beans, and quinoa, much earlier. Llamas were domesticated by at least 2500 B.C.E. Maize agriculture probably spread south from Central America to the highlands, then to the arid Peruvian coast by 800 B.C.E., where maize was grown in large-scale irrigation schemes in river valleys. This intensive agriculture was the staple of the coastal civilizations that developed by the Pacific after 800 B.C.E.

Maize agriculture spread into the North American Southwest by about 1500 B.C.E., where cold winters and arid conditions made cereal agriculture difficult. Nevertheless, maize and bean agriculture became the basis of the sedentary Pueblo Indian cultures that developed in the Southwest after 2,000 years ago. The well-known Anasazi, Hohokam, and Mogollon cultural traditions of the Southwest were the ancestral foundations of modern Southwest Indian society. Between the 10th and 13th centuries C.E., some Anasazi pueblos, notably in Chaco Canyon, New Mexico, and Mesa Verde, Colorado, housed hundreds of people, especially during important seasonal ceremonials. Southwestern Pueblo societies were successful, highly flexible adaptations to unpredictable, semiarid environments. As such, they never achieved the degree of social complexity found further east in North America.

By 2500 B.C.E., hunter-gatherer societies in the eastern woodlands of North America were planting native plants on a regular basis. Maize crossed the southern Plains into eastern North America during the first millennium C.E., but did not become well established until after 800 C.E. By this time, many eastern societies were living in sizable, sedentary communities, presided over by powerful kin leaders.

LATER OLD WORLD PREHISTORY (3000 B.C.E. AND AFTERWARD)

State-Organized Societies

Within 5,000 years of the appearance of farming villages in the Near East, state-organized societies developed in Egypt and Mesopotamia. Subsequently, other preindustrial civilizations of great complexity emerged—not only in the Near East, but also in south Asia, Southeast Asia, and China. They also flourished in Central America and the Andean region of South America. Such state-organized societies operated on a large scale with centralized political and social organization.

They were marked by class stratification, intensive agriculture, and were based on assumptions of social inequality. All had complex government bureaucracies and were often ruled by despotic leaders who governed as divine monarchs.

State formation was not a universal phenomenon. Many archaeologists theorize that states emerged because they were beneficial as a way of organizing both food supplies increased through intensified agriculture and trade, and external relations with neighbors. They believe states emerged in environmental settings with severe population problems or shortages of agricultural land. Effective, centralized management of trade monopolies, and of food production, through state-organized irrigation systems and other means, could bring ecological imbalance under control. For example, both Ancient Egyptian pharaohs and Aztec rulers in Mexico employed hundreds of officials to ensure that all available land was cultivated efficiently.

Webs of Relations

For all their complexity and sophistication, the early civilizations ruled over relatively small geographical areas by modern standards. But their insatiable demands for exotic raw materials—for gold, copper, iron ore, even such prosaic items as timber or textiles—brought them into contact with dozens of prehistoric, nonliterate societies that lived on the margins of, or outside, their boundaries. The last 5,000 years of prehistory are remarkable for the ever-expanding tentacles of interconnectedness that linked hundreds of prehistoric societies with one another, and with more complex literate civilizations many miles away. This web of relations began to expand as early as 7000 B.C.E. in the Near East, when the obsidian trade linked small farming settlements seeking fine toolmaking stone. In time, these small regional trading networks became well-traveled caravan routes that linked Mesopotamia with the Mediterranean and the Nile Valley to the Levant.

During the fourth millennium B.C.E., long-distance trade exploded throughout the Near East, linking societies all the way from the Indus Valley in Pakistan and Iran with

Mesopotamia, the Levant, Anatolia, and the Nile Valley. This rapidly evolving "world system" transformed human life. A millennium later, it embraced not only the Near East, but Cyprus, the Aegean, and mainland Greece as well. It developed because of an insatiable demand for nonlocal raw materials in different ecological regions where societies were developing along very similar lines.

The Sumerians of Mesopotamia, for example, lived in a largely treeless environment with no metals. Yet they produced large grain surpluses, which they traded for timber and metals with the highlands by boat and across the desert by donkey camels. Nowhere did this emerging world economic system have a greater impact than in Africa.

INDIAN AGRICULTURE: PEDESTAL OF OUR ECONOMY

In his historic "Tryst with Destiny" speech Pt. Jawahar Lal Nehru said,"Everything can wait, but not agriculture."

Though Indian economy is growing by leaps and bounds, agriculture still remains its Achilles heel. There is a wide disparity between the producer and consumer. Farming has largely become an un-remunerative occupation. However success we may achieve, the ground reality remains that India has failed to bring its farmers to the mainstream. Sad to say, but we have become oblivious of our villages. Can we expect a booming economy without inclusion of agriculture? In fact, this is one of the reasons why India is still a developing nation. The condition has worsened to such an extent that nobody likes to pursue farming.The NSSO estimates that 40% farmers would like to quit farming if they had the option to do so.

There are a plethora of reasons for this widespread deprivation and disenchantment. Low public investment, inadequate irrigation facilities, fragmented land holdings, low technology application and farmers debt have spelt doom for this sector that is supposed to be the lifeline of our economy. Prime Minister Manmohan Singh acknowledges that a double digit growth would forever elude India without a dramatic turnaroud of the rural landscape. Govt investment in agriculture has fallen from 14.9% in the first five-year plan to 5.2% in the current plan. Whatever substantial is allotted for the farmers becomes the morsel of our corrupt politicians. One of the reasons for backwardness of Indian agriculture is its dependence on monsoons which are largely erratic in nature. Currently only 40% of the land under cultivation is irrigated and a large number of river-valley projects have failed to meet their deadlines. The problem of fragmented land–holdings makes it inept for the use of modern machines and techniques.

The year 2006 was agog with news of farmers suicides in various parts of the country. Farm input costs continue to rise but the output prices remain stagnant; debts are mounting but support systems are nowhere to be seen.Lack of institutional credit is the major cause of rising debts among farmers. As a result marginal farmers turn to private money-lenders for loans at interest rates anywhere between 40 to 120% p.a. to purchase seeds, fetilizers and other agricultural inputs. M.S Swaminathan's recommendation that India needs to bridge the knowledge deficit in its farm sector is well received by the planning and policy circles.There is an urgent need for investment in R&D in agriculture. India spends 0.31% of its GDP on R&D which is far below industrialized nations. Agricultural institutions like ICAR and CRRI should be revamped according to present day needs. The UPA govt's current aim is to raise the agricultural growth rate to 4% from a worrisome 1%.It's not that Govt has been a mute spectator all these years. Whatever self-sufficiency we have achieved in food-grain production is because of the prudent measure by past Govts. India's stand at WTO is a matter of pride for all developing countries of the world.

Indian agriculture is in a quandary. The acquisition of farm land for industrial purposes has opened can of worms for farmers. The ideal step would be use of barren land, proper compensation,&industry-farmer partnership. India certainly needs a second green Revolution and to usher that we will have to look back at the first green revolution model which involved collaboration between central and state governments, agricultural universities and research institutions.

ALTERNATIVE STRATEGIES FOR INDIA'S DEVELOPMENT AGRICULTURE, FOOD SECURITY AND HUNGER

Scenario 1: *It was too late. By the time, Jai Lal, a landless agricultural worker of Bandali village, in Sheopur district of Madhya Pradesh, in the heartland of India, returned to share the good news with his wife – that he finally managed to get a petty job with a shopkeeper – she had succumbed to hunger. A week later, graves were dug for his two children, both unable to continue with the prolonged fight against hunger.*

Jai Lal's family paid a heavy price for the faulty agricultural policies that are being relentlessly promoted and pushed in the name of economic growth and development. Jai Lal is not the only victim of a development paradigm that turns a blind eye to the resulting human suffering. Travelling around the country, I am no longer shocked at the plight of the rural masses, who unknowingly continue to pay a heavy price for the faulty agrarian policy thrust upon them. What hurts me is to see that even fifty-five years after Independence, growing hunger and inequalities do not prick the conscious of the nation. Moreover, in the era of free markets and ruthless competition, human compassion has become the biggest casualty.

There is no other plausible reason that can explain why Jai Lal lost his family. After all, Jai Lal's family died of hunger when more than 45 million tonnes of foodgrains were stacked in the open, much of it rotting for want of adequate storage facilities. This was in early 2003. Two years earlier, the country had a record 65 million tonnes of food surplus, at a time when nearly 320 million – a third of the world's estimated 840 million hungry – looked in disbelief at the mountains of the food stocks that lay decaying in front of their dry eyes. In fact, hunger and poverty have proved to be robustly sustainable.

While people die of hunger, the government sits atop a mountain of food grains. In 2001, starvation deaths were reported in over 13 states while the storage facilities of the Food Corporation of India (FCI) - were full of grains, some of it rotting and rat-infested. There was a proposal to dump it in the sea, to make storage space for the next crop, when export markets could not be found for this surplus. Such was the quantity of food kept in the open, that if each bag was stacked one upon the other, there was no need to launch a scientific expedition to put a man on the moon. You could simply walk to the moon and come back.

A report of the Standing Committee of Parliament estimated that the government was spending Rs 62,000 million every year to maintain these food stocks. Mainline economists and agricultural scientists did not question the necessity of maintaining the surplus stocks when millions were sleeping with empty stomach. Instead of making the surplus grain available to the poor, nearly 24 million tonnes from the unmanageable food surplus was diverted for exports in 2002-03, and that too at a price that was actually meant for people living below the poverty line. Another six million tonnes were released for the open trade at the same price. The poor and hungry however were not the beneficiary. They are expected to live on hope.

Despite the Planning Commission pulling down the percentage of poor and poverty stricken from its unread documents, the magic trick of playing with numbers hasn't made any difference to the growing disparities. Amidst recurring political elections, and the brazen marketing hype to sell images of growth and development, the shameful paradox of hunger at times of plenty has been quietly buried under heaps of grain that continue to rot in the open. That 75 lakh people, more than the population of Switzerland, had applied for a mere 28,000 lowly-paid jobs in the Indian Railways (in Nov 2003), is no longer a matter of concern at times when the country is on a fast track information highway. Not to discount the achievements in information technology, the fact remains that IT has provided only five lakh job opportunities.

Scenario 2: *Far away, Argentina, the world's fourth biggest exporter of food, faces an unprecedented socio-economic crisis. As the vast, fertile country continues to increase exports of meat, wheat, corn and soya this year, a catastrophe has hit the under-privileged in the countryside. As hunger multiplies, images of stunted, emaciated children have scandalised Argentina, long known as the grain store of the world.*

Perceived by the neo-classic pundits as the glorious model of economic growth, an unprecedented humanitarian crisis confronts Argentina. The Guardian (Nov 25, 2002) explains the dichotomy of economic growth in Argentina, quoting the Centre for Child Nutrition Studies, which advises the World Health Organisation, as saying that 20 per cent of children in the Latin American country are suffering from malnutrition. Dr Oscar Hillal, the deputy director of the children's hospital in Tucuman, said: "This is not Africa, this is Argentina, where there are 50 million cattle and 39 million people - but where we have a government which is totally out of touch with the people's needs."

Some of the children pictured in northeastern Tucuman province had bloated stomachs, blotchy skin and dry hair associated with severe protein deficiency. The national charity Red Solidaria said that 60 children a month were being taken to hospital with severe malnutrition, and 400 were being treated as outpatients. Five non-government organisations from Tucuman province had filed a legal suit against Tucuman's governor for "wilful neglect" of the children who have died of malnutrition in his province, where 64% of people live in extreme poverty. They accused him of diverting national funding for social programmes into "clientelism and corruption".

In India too, with the increased domination of market forces in the food sector, and reduced public policy intervention for food security, food prices have increased. Government has promised to further cut down subsidies and reduce the government's intervention in foodgrain procurement. Already, India's Export-Import Policy of 2001-02 lays a major thrust on promotion of agricultural exports, the third largest global producer of food. Meanwhile, exports have increased by 10 per cent every year since 1991. They rose from Rs 29.7 billion in 1994 to Rs 76.7 billion in 1997.

Scenario 3: *The recent spate of farmers suicides that began in Andhra Pradesh, and caught the national attention in the aftermath of the stunning electoral verdict that unseated the BJP-led government at the electoral hustings in 2004, has been baffling. Ever since the new government in Andhra Pradesh was sworn-in on May 14, 2004 more than 600 farmers have committed suicides. This was the death toll in the suicides register till July 15.*

The situation in several other states, including the frontline agriculture states of Punjab and Haryana, is no better. In the Vidharba region of Maharashtra, for instance, more than 200 farmers have committed suicide during the year.[2] Karnataka, West Bengal, Orissa, Tamil Nadu, erala, Madhya Pradesh, Rajasthan, and Uttar Pradesh are all faced with the unexplained phenomenon of farmers taking their own lives.[3]. What has been shocking is that the spate of been shows no signs of ending even after short and long-term relief packages aimed at relieving farmer's misery were announced.

Although the newly-elected government of Andhra Pradesh (and followed closely by Tamilnadu) have moved in quickly by announcing free power to farmers,[4] what is more depressing is that the governments are clueless of the reasons that forces farmers to commit suicide. The reason is obvious. No one has the political courage to point a finger at the clear verdict against the industrial farming model being forced down the throat of small and marginal farmers.[5]

With the high-chemical input based technology that mined the soils and ultimately led to the lands gasping for breath, with the water-guzzling crops (hybrids and Bt cotton) sucking the groundwater acquifer dry, and with the failure of the markets to rescue the farmers from a collapse of the farming systems, the tragedy is that the human cost is entirely being borne by the farmers.[6] In Punjab, for instance, of the 138 development blocks, 84 have already been declared dark zones, the level of groundwater exploitation in these blocks has been in excess of 98 per cent against the critical limit of 80 per cent. Six of the 12 districts in the State have recorded groundwater utilization rate of 100 per cent. The National Bureau of Soil Survey and Land Use Planning in India estimates that nearly 120 million hectares of the total cultivable land of 142 million hectares in the country is degraded. The fundamental issue of destruction of sustainable livelihoods is not at all being addressed.[7]

Scenario 4: *At first impression, news reports that appeared in 2002 the American media looked like emanating from a drought-stricken village in India's hinterland. Till of course you see the dateline. You continue to read in utter disbelief. About 100 desperate farmers and rural residents praying for rain at the St. Patrick parish church in Grand Rapids, Ohio. With hands clasped and eyes cast downward, they seek divine intervention. "None of us have control over whether it is going to rain or not," said Sister Christine Pratt, rural life director for the Catholic Diocese of nearby Toledo told Reuters, the wire agency. "But the people are praying for one another and there is some hope."*

Another report in the Washington Post stated President George Bush did not extend finances under drought relief in addition to the support that came from $180 billion farm bill he signed in May 2002. The president however underscored his commitment to helping farmers under current programs, including the Agriculture Department's decision that provides $150 million in surplus milk — "spoiled milk," as Democrats called it — to be made available for use in animal feed in four drought-stricken states, including South Dakota.

Cattle were dying and crops shrivelled. Fodder become scarce, and, therefore, the need to feed surplus 'milk' instead. There was a scramble for new water sources as town and city residents were asked to stop watering lawns and washing cars. In heat-baked fields ranchers sold off herds rather than let them starve for lack of pasture. "I have never seen it like this and I'm 60 years old," said Richard Traylor, who owns 37,000 acres in Texas and New Mexico but had sold off much of his cattle herd.

Serious hydrological problems with wells and reservoirs emerged. Streams went dry. The groundwater table fell drastically. Wildfires became more rampant, and an estimated 4.6 million acres, had been scorched, twice the average acreage burnt in the previous decade. "It is pretty dire," Mark Svoboda, climatologist for the National Drought Mitigation Center was quoted as saying. From southern California to South Carolina and from Montana to New Mexico, individuals and industries were suffering, the news agencies reported.[8]

America was faced with its worst drought since the days of the great 'dust bowl' of the 1930s. By a strange coincidence, far away, India too was reeling under its worst drought of the century. As many as 26 of the 50 American States were reeling under a severe drought, with "exceptional drought" conditions — the worst level of drought measured — prevailing in thirteen states, including New Mexico, Arizona, Colorado and Utah. In India, drought had ravaged through 12 of the 28 states. Such was the crop damage that like the drastic reduction in foodgrain production in 2002-03 in India, the US wheat production too was anticipated to fall to its lowest levels in nearly 30 years.

Scenario 5: *At every national and international conference, it is not unusual to see slide projections that point at the low productivity in India and for that matter in other developing countries. The projections for area and productivity under cereals, including wheat and rice, and crops like sugarcane, cotton and vegetables points to the prevailing dichotomy. India ranks among the top five countries (often among the first two) having the largest area under crops like wheat, rice, cotton, sugarcane, and vegetables. India's ranking in productivity brings it to the bottom of the chart, with per hectare yield or productivity hovering amongst the lowest five or ten countries. The conclusion therefore is that increasing productivity will bring more income to farmers and thereby increase their presence and competitiveness in the international market.*

Farmers are being misled to believe that diversification from staple grains to cash crops is the only way out to escape an uncertain future. At the same time, farmers are being asked to increase crop productivity to remain competitive in an era of 'free' trade. Since the global trade parameters are being relaxed and phased out, increasing productivity is being touted as the new survival mantra. The high productivity refrain comes in handy for the biotechnology industry to bring in expensive and risky technologies thereby further compounding farmers' woes. In the bargain, it is the farmer who faces the brunt, often opting for the fatal route to escape the humiliation and distress that such half-baked advice brings in.

Take the case of rice, the most important staple food crop of India. In the year 2000, India's rice paddy yield was hovering at 3008 kgs per hectare. In Thailand, the major rice exporter, paddy productivity stands at 2329 kgs. In the United States, the average yield per hectare was more than double at 7037 kgs. If productivity alone was the criteria, the US should have captured the entire world market in rice. And at the same time, Thailand shouldn't have been able to export rice considering that its average productivity is lower than even India.

Moreover, even with such low rice paddy productivity, India had a record procurement of 20.9 million tones of rice in the 2001-02 marketing season. The grain stock build-up over the last few years has seen India's rice and wheat surplus increase to an unmanageable level of 51.4 million tones in October 2002 (against a record 65 million tones in June 2001). In fact, chief ministers of surplus rice producing states of Punjab, Haryana, and Andhra Pradesh have been repeatedly asking their farmers not to produce more of rice as they have no place to stock it. The Central government too has been toying with the idea of getting out of food procurement leaving farmers to the vagaries of the market.

In the United States, however, despite the high paddy productivity, farmers find its cultivation uneconomical. The US government, therefore, continues to subsidise the American farmers. Estimates point that the American farmers receive an average subsidy of US $ 30,000 per farm per year. As if this is not enough, the new Farm Bill brings an additional federal support of US $ 180 billion for the next ten years. If high productivity is the criteria for global competitiveness, there is no plausible reason why the American farmers would depend upon government doles for survival.

To ask the Indian farmers, therefore, to increase paddy productivity is to merely push them into a death trap. Already, rice farmers in Punjab and Andhra Pradesh continue to suffer for producing more. For the past two years, with the Food Corporation of India (FCI) refusing to buy paddy under one pretext or the other, distress sale has become a common phenomenon. At many a places, a number of rice farmers preferred to commit suicide waiting endlessly for buyers in the markets. The scenario for wheat producers is no different. They too are faced with the 'produce and perish' syndrome.

1. Keynote address at the national seminar on "Alternative Strategies for Development" held at the RCVP Noronha Academy of Administration and Management, Bhopal, India, from Aug 10-12, 2004.
2. Supriya Menon reports "Every Week in Maharashtra, One Suicide", Tehelka, July 10, 2004
3. More details are provided in an excellent cover story in Frontline, July 2, 2004 "An Agrarian Tragedy"
4. World Bank has expressed its displeasure over the move to supply free power to farmers.
5. Devinder Sharma in New Scientist, July 8, 2000: "Green Revolution turns sour"
6. KPA Prabhakaran Nair has brought out the technology aspect of the farmers suicides in his essay "When Farm technology falls short of expectations", Business Line, July 6, 2004.
7. See Devinder Sharma's "My Vision for a Global Agriculture", a talk delivered at the UK Food Group's dialogue on agricultural trade reforms, subsidies and future of small and family farms and farmers, at London, June 30, 2004.
8. There isn't a time when an educated Indian doesn't search for answers from "America—the dream land" for the problems that crop up time and again back home. The solutions to India's crisis on the farm front, whether it pertains to sustainability, role of markets, or the recurring drought — rest in the way America has managed its crop lands. After all, the United States has put together a drought-mitigation strategy, which many

feel India needs to follow immediately. The author had earlier analysed the faulty agriculture model that is being imposed onto India.

Every day, some 5,000 children somewhere in India succumb to diseases, related to malnutrition and hunger. Everyday, thousands of rural people – without land and adequate livelihoods – constituting a reservoir of frustration and disaffection, trudge to the cities, their abject poverty contrasting vividly with the affluence of the urban centres.

In the remote Kalahandi district of Orissa, synonymous with drought, hunger and misery, starvation exists amid plenty. Millions of hapless, deprived and landless wait endlessly for the rice they producc. The crop they harvest finds its way to the food reserves of the nation, much of it being devoured by rats or exported. Kalahandi is among Orissa's biggest contributors of foodgrains to the FCI.

The paradox of plenty, no longer, is confined to the terrains of Kalahandi. A nation burdened with the guilt of one Kalahandi, refuses to even look for numerous other pockets of hunger and destitution. India too is meanwhile faced with a Kalahandi syndrome – foodstocks piling up at a time when a third of the world's 840 million hungry and poor, living in India, do not have the means to purchase it. In addition, the resulting damage to the resource-base, on which were laid the strong foundations of the traditions of agriculture, have been ruinous and threaten the survival of the nation.

For a country, which emerged from the throes of a 'ship-to-mouth' existence, to be subsequently able to build up foodgrain reserves from homegrown wheat and rice, sustainable agriculture was the unmistaken path to equitable growth, development, and national food security. The green revolution technology, which ushered in 'food self-sufficiency', however came with enormous environmental costs. It used massive amounts of chemicals, fossil fuels and water. In energy terms, it was less efficient than many traditional farming systems. Monoculture, mechanical ploughing, soil erosion, the extension of crops into forests and the use and abuse of chemicals has contributed to the second-generation environmental impacts that the intensively-farmed lands of the country are grappling with.

Fertilisers and pesticides were aggressively promoted, with huge subsidies being doled out to keep the fertiliser companies afloat, without realising the resulting devastation these chemical inputs have wrought on the sustainability of agriculture. At no stage, did the scientists call for a mid-term correction to rectify the imbalance and destruction of the soil fertility through excessive application of the chemicals (see box 1). The second-generation environmental impacts became so serious that the Consultative Group on International Agricultural Research (CGIAR), which governs the 16 international agricultural centres, did launch an initiative for studying the negative impact of the green revolution model on sustainability of agriculture in the Indo-Gangetic plains but the results were never made public.

Pesticides are a Waste of Time and Money

Three decades after the launch of the Green Revolution, agricultural scientists are now discovering that chemical pesticides are a complete waste of time and money. They have realized the grave mistake only after poisoning the lands, contaminating the ground water, polluting the environment, and killing thousands of farmers and farm workers.

Says an IRRI press release (July 28, 2004): Imagine 2,000 poor rice farmers in Bangladesh, whose average farm income is around US$100 per year, suddenly take on the role of agricultural scientist. Over the course of 2 years — 4 seasons – they prove that insecticides are a complete waste of time and money. IRRI senior entomologist Gary C. Jahn, states: "To my surprise when people stopped spraying, yields didn't drop — and this was across 600 fields in two different districts over 4 seasons. I'm convinced that the vast majority of insecticides that rice farmers use are a complete waste of time and money."

This is the outcome of a joint IRRI-British DFID's Livelihood Improvement Through Ecology (LITE) project, which has demonstrated that insecticide can be eliminated and nitrogen fertilizer (urea) applications reduced without lowering yields. "We've reduced insecticide use among participating farmers by 99%, and by 90% among non-participating farmers in the same villages, Dr Jahn added.

What's more, if LITE continues as it has started, in less than a decade, most of Bangladesh's 11.8 million rice farmers — almost a 12th of the country's population of 141 million, according to the Bangladesh Rice Research Institute, a key project partner — will have eliminated insecticides and optimized their fertilizer use.

Similar studies in the Central Luzon province of the Philippines and in certain parts of Vietnam have already demonstrated that pesticides were not required. Does it not mean that agricultural scientists had pushed and promoted chemical all these years without looking for viable and sustainable alternatives? Does it not mean that the technology for productivity increase was not based on sound ecological and environmental parameters? Does it not mean that the land grant system of research had ignored the potential of agriculture growth that existed in the developing countries, based on time-tested technologies and sustainable farming system?

If it has taken 30 years to realize that the technology promoted by the USAID and blindly aped by the National Agricultural Research Systems in the developing countries, and that too after inflicting an irreparable damage to human health and environment, was faulty; what is the guarantee that genetically engineered crops will not leave behind still more damaging consequences? Who will be responsible for the destruction that is being enforced through genetic manipulations?

For several years now, almost two decades, basking in the afterglow of the green revolution, and with abundant monsoons to boost it, farming and agriculture ceased to attract attention. Policy-makers began to believe that there was no cause for undue concern since the country had the capability to 'reproduce' another green revolution. Not realising that green revolution had run out of steam, and rural despair was growing. Slow on-farm agricultural employment and the overal employment growth trailing behind the gorwth of the labour force, more and more people began migrating to the cities.

The alarm bells have been ringing for quite some time. The spectacular yield growth recorded in the post-Green Revolution years in Punjab and Haryana have receded into history. Among the multiplicity of problems confronting agriculture, rapid fragmentation of land holdings is keeping pace with increasing population. In 1976-77, the average size of the holdings was estimated at two hectares, and in 1980-81, it came down to 1.8 hectares. Today, it stands at a mere 0.2 hectares. The total number of land holdings in 1981 were around 89 million, today these have crossed 100 million.

Thirty years after the dawn of green revolution, Indian farmers realised that their love affair with intensive agriculture was on decline. Despite a bountiful monsoon (14 normal monsoons in a row), harvests were not as plentiful as could have been expected. As intensive farming began to bare its fangs, mining the ground water, and destroying the soil fertility, sustainable livelihoods began to fall apart. At the same time, by the turn of the century, per capita foodgrain availability had dropped to an abysmal low of 152 Kg, nearly 23 kgs less than early nineties. This compared favourably with the stark hunger that prevailed in sub-Saharan Africa, and was no better than the crisis-laden food situation that existed at the time of the Bengal Famine.

Green revolution had not only gone sour, it has now turned red. The unexplained number of huge number of farmer suicides is a testimony to the entire equation going wrong.

The philosophy of agricultural planning is changing. Gone are the days when the nation's emphasis was solely on attaining self-sufficiency in foodgrain production. Gone are the days when farmers were the newly Independent India's heroes, revered for their role in keeping hunger and sure starvation at bay. Today, at a time when food production struggles to barely keep pace with the burgeoning population growth, farmers are being asked to diversify, produce crops that are suitable for export and to compete in the international market. With promise of cheap food available off the shelf in the global market, the focus has shifted from agriculture to industry, trade and commerce, from the small and marginal farmers to the agri-processing companies, which alone can bring in investments and add value to produce.

Whither second Green Revolution: The reforms being introduced in the name of increasing food production and minimising the price risks that the farmers continue to be faced with, is actually aimed at helping the agribusiness industry. Whether it destroys the production capacity of the farm lands and leads to further marginalisation of the farming communities does not figure in the policy planning process. Encouraging contract farming, future trading in agriculture commodities, land leasing, forming land-sharing companies, allotment of homestead-cum-garden plots, direct procurement of farm commodities and setting up of special purchase centres will however drive out a majority of the 600 million farmers out of agriculture.

In a country where land holdings are meagre, the biggest challenge is to ensure how can agriculture be made more attractive for these small and marginal farmers. At the same time, in the green revolution areas, comprising Punjab, Haryana, western Uttar Pradesh, parts of Andhra Pradesh, Tamil Nadu and Karnataka, agriculture faces a severe crisis in sustainability. As a result, Punjab and Haryana are fast heading towards desertification – a process that leads to the inability of the lands to sustain the production levels achieved at the height of the green revolution era.

Although the land holding size is diminishing, the answer does not lie in allowing the private companies to move in by way of contract farming. Private companies enter agriculture with the specific objective of garnering more profits from the same piece of land. These companies, if the global experience is any indication, bank upon still more intensive farming practices, drain the soil of nutrients and suck ground water in a couple of years, and render the fertile lands almost barren after four to five years. The once fertile and verdant landscape will fast turn grey. These companies would then hand over the barren and unproductive land to the farmers who leased them, and would move to another fertile piece of land.

Rebuilding ground water resource should be an essential parameter for any meaningful agriculture reforms. Unfortunately, at a time when excessive withdrawals of underground water have already become a major political issue, cropping pattern continues to play havoc with the irrigation potential. The lessons from the other contract farming models should be too apparent. Sugarcane farmers, who follow a system on cane bonding with the mills, actually were drawing 240 cm of water every year, which is two and a half times more than what wheat and rice requires each on an average. Rose cultivation that was introduced in Karnataka a few years back, required 212 inches of groundwater consumption in every hectare. Contract farming will therefore further exploit whatever remains of the ground water resources.

Legal recognition of land leasing is therefore no protection to farmers. Once the production capacity of the land has been destroyed what can the farmer be expected to reap thereafter. Knowing this, the government is talking of homestead-cum-garden plots for those who lease out their lands. The objective is simple: to pacify those who question the impact of contract farming on household food security. Policy makers and planners are not even aware of the basic objective behind encouraging contract farming. Often it is said that these companies will only be there for helping the farmers in marketing. What is deliberately not being mentioned is that nowhere in the world are private companies involved with contract farming just to help the farmers find a marketing outlet.

Punjab, Andhra Pradesh and subsequently other state's foray with contract farming therefore is a misplaced adventure.

It is actually accentuating the sustainability crisis on the farm front by destroying whatever remains of the farmland's production capacity with more intensive and destructive farming systems. The resulting monoculture also destroys the agriculture biodiversity in the region thereby hitting sustainability parameters. In simple words, contract farming is the modern version of the 'slash and burn' agriculture (*jhum* cultivation) that the tribals followed in the northeast parts of the country. Tribals were doing it for environmental reason, whereas the private industries are forcing this for commercial motive alone.

Already contract farming has done irreparable damage to agriculture in countries like the Philippines, Zimbabwe, Argentina and Mexico.

Allowing direct procurement of farm commodities, setting up special markets for the private companies to mop up the produce, and to set up land share companies, are all directed at the uncontrolled entry of the multinational corporations in the farm sector. Coupled with the introduction of the genetically modified crops, and the unlimited credit support for the agribusiness companies, the focus is to strengthen the ability of the companies to take over the food chain. Significantly, the state governments have opposed the agriculture reforms, terming it as a recipe for the entry of multinational corporations in agriculture.

Agribusiness companies in reality hate farmers. Nowhere in the world have they worked in tandem with farmers. Even in North America and Europe, agribusiness companies have pushed farmers out of agriculture. As a result, only 9,00,000 farming families are left on the farm in the United States. In the 15 countries of the former European Union, the number of farmers has come down to less than 7 million. The underlying message is crystal clear: farmers should get out of agriculture. In India, the same prescription will lead to an unforeseen catastrophe, worsening food insecurity and multiplying hunger.

To expect farmers to collectively mobilise the land resources to facilitate access to modern technology and professional management in the farm sector, a concept being floated in the name of land sharing companies, too is aimed at private control of the farmland. In India, except for a handful of such cases, farmers do not have the ability to pool land resources unless backed by a private company. In other words, land sharing is another name for contract farming. All such experiments would be forcing the farmers to shift from staple foods to cash crops like cut flowers, tomato, strawberries, melons which do not meet the food security needs at the macro level. At the same time, the intensive nature of cash crop cultivation, requiring more external inputs, would do more damage to the environment.

Agriculture Reforms – The Way Ahead

Sustainable agriculture thus sustains rural livelihoods. This in turn is directly linked to the nation's as well as the household food security. Any development alternative to ensure long-term food security therefore has to be linked to sustainable agriculture.

Let me therefore draw the outline of the sustainable farming systems that the country needs to focus on. This is the overall framework under which location-specific alterations and adaptations need to be tried.

What is needed is a fresh approach that takes the ground realities into consideration before embarking upon any policy imperatives. I am trying to make an attempt, presenting a collection of five of the important rational decisions, which would certainly initiate the revival of Indian agriculture:

Sustainable farming: Indian agriculture faces an unprecedented crisis in sustainability. Foodgrain productivity in the food bowl, comprising Punjab, Haryana, and western Uttar Pradesh, is on the decline. The green revolution areas are encountering serious bottlenecks to growth and productivity. The dryland areas (comprising nearly 70 per cent of the cultivable lands) continue to drown in misery and apathy. Excessive mining of soil nutrients and groundwater have already brought in soil sickness. Indiscriminate use of chemical pesticides has done serious harm to environment, human health and ecology. Introducing new Centrally Sponsored Schemes or contract farming to improve production in these areas is going to be counter-productive. Banking upon genetically engineered crops to take care of the second-generation environmental impacts is sure to worsen the existing crisis. Outlays earmarked for genetic engineering in agriculture also need to be diverted to sustainable agricultural practices.

Encouraging sustainable and traditional farming practices therefore is the only way ahead. Agricultural research must reorient itself to meet the new challenges resulting from the collapse of the green revolution technology. Investments and increased outlays for agricultural research that is based on external chemical inputs like fertiliser and pesticides need to be discouraged. Instead, financial allocation should bc made for reviving low-input agriculture, which uses cheap and locally available technology and in turn improves production and protects environment. This has been amply demonstrated in several parts of the world (see the accompanying box). Water productivity and efficiency has to be the hallmark of agricultural research based on the local conditions.

Local Solutions: For the past three decades, more so after the introduction of the land-grant system of education, the focus is on finding global solutions to local problems in agriculture. The World Bank/IMF, the Consultative Group on International Agricultural Research (CGIAR) and now some of the major donors like DFID and GTZ have been embarking of translocating alien approaches to agricultural improvement and have thereby exacerbated the crisis on the farm front. The Indian Council for Agricultural research (ICAR) too has blindly followed the land grant system of research and education, the negative results of which are now becoming apparent. Ignoring the traditional knowledge and time-tested technologies has created a crisis on the farm front. This process must be immediately stopped, if not reversed. Given the diversity of the agro-ecological regions, sustainable agriculture needs location-specific solutions.

International agricultural research, as well as the national agricultural research systems, should re-orient the focus of farm research based on the principles of – farmer friendly, environment friendly and long-term sustainability. Instead of the 'Lab-to-Land" approach, which has done immense damage to agriculture globally, the emphasis should be on learning from the land, meaning going back to farmers and the traditional farming systems. Technology need not always be high-tech and sophisticated.

It can be simple and effective. This can only be ensured if the effort is to fit the new and improved technology to farmers need rather than asking farmers to fit into the technology package developed. This can only happen if farm research is brought back to the public sector. All technology should be freely available, and should not come with any proprietary tags.

Dryland farming: Despite the former Prime Minister Mrs Indira Gandhi's emphasis on dryland farming, agricultural scientists as well as the policy makers have failed the dryland farmers. This is essentially because the entire thrust of dryland research was to bring in an external model in which the dryland farmer, who manages to survive against all odds, would fit in. No effort was made to improve the existing technology base under numerous location-technology specifications.

At the same time, drylands continue to be plagued with recurring drought engulfing vast tracts of central and north-western India. The increased emphasis on water harvesting notwithstanding, the reduced availability of water is emerging as a major social and economic crisis.

This is because much of the investment is going into a faulty technology of rainwater harvesting, called the "Ridge to valley" system, a technology imported from the United States. In addition, the cropping pattern has to be evolved keeping in mind the water availability.

At present, more the water requirement for hybrid crop varieties more is its cultivation in the water-scarce regions. This is scandalous and unless the cropping pattern is rectified no measures to protect and preserve water resources will be effective.

Improved Crops Mine Water

High-chemical input based technology has already mined the soils and ultimately led to the lands gasping for breath, with the water-guzzling crops (hybrids and Bt cotton) sucking the groundwater acquifer dry, and with the failure of the markets to rescue the farmers from a collapse of the farming systems, the tragedy is that the human cost is entirely being borne by the farmers. Green revolution was projected to have saved the country some 58 million hectares of additional land to be brought under the plough to produce more food, whereas almost twice that land mass has been rendered degraded and ecologically devastated in varying degrees in its aftermath.

Green revolution has not only gone sour, it has collapsed. The unexplained number of huge number of suicides a testimony to the entire equation going wrong. However, the fundamental issue of destruction of sustainable livelihoods is not at all being addressed.

All these years, for instance, the dryland regions of the country, which comprise nearly 75 per cent of the total cultivable area, have increasingly come under the hybrid crop varieties. While the crop yields from the hybrid varieties was surely high, the flip side of these varieties – these varieties are water guzzlers – was very conveniently ignored. For the sake of comparison, let us take the example of rice.

Not only rice hybrids, all kind of hybrid varieties that require higher doses of water – whether it is of sorghum, maize, cotton, bajra, and vegetables are promoted in the dryland regions. In addition, agricultural scientists have misled the farmers by saying that the dryland regions were hungry for chemical fertilisers. The harmful combination of chemical inputs with water guzzling crops have played havoc with the drylands turning the lands not only further unproductive but also barren. The water table plummeted, the impact of deficient rainfall be-

came more pronounced forcing farmers to abandon agriculture and migrate. As if this was not enough, Bt cotton requiring more water than hybrid cotton, was knowingly promoted so as to allow the seed industry to make profits.

Investments in rainwater harvesting need to be immediately shifted to the revival of the traditional forms of water conservation – ponds and tanks. Fodder cultivation, crop planning according to the water needs and availability and the emphasis on the local breed of cattle (and improving its productivity, rather than importing exotic breeds) need to be encouraged. Dryland crops, and that include coarse cereals, pulses and oilseeds require adequate policy measures that bring shine to these forgotten grains.

Farmers in the rain-fed areas also need to be insured against drought. This can be ensured by making it mandatory for the foreign insurance companies to invest at least 40 per cent of their funds for farm insurance.

Sugar mills: Sugarcane is the biggest threat to India's food security. The unprecedented addition of new sugar mills by successive governments has created a major crisis on the agriculture front. Requiring good fertile and irrigated land for cultivation, its growth is at the cost of staple foods like wheat and rice. With the per hectare productivity of foodgrains on the decline in the frontline agricultural states, diversion of good fertile land to sugarcane is not without accompanying hiccups. What makes the switchover to sugarcane a pernicious trend is its enormous water requirement. Sugarcane, in fact, is the biggest threat to India's food security.

Since there is no shortage of sugar in the country, and with a large number of mills actually being rendered unviable over the past two decades, an immediate ban needs to be imposed on setting up any new sugar mill. All Budgetary support to the sugar industry needs to be withdrawn as it has led to a serious environmental crisis. Reduce the area under sugarcane, improve productivity, disband most of the unproductive sugar mills, and give a new lease of life to the cane areas.

Instead the focus should shift to pulses and fodder crops. Pulses are essential for country's nutritional security and fit very well into the harsh environments. Sugarcane growers in most parts of the country can easily be made to shift to pulses cultivation given the right incentive. Such a renewed emphasis will not only help farmers and consumers alike but also rejuvenate the environment and help in restoring soil health. Pulses have the inbuilt capacity to draw nitrogen from the atmosphere.

Marketing: Providing an assured and remunerative market for agricultural producers cannot be left to the market forces. The food policy imperatives of public distribution system and announcing the procurement prices before the crop season have to be further strengthened. Agri-processing too needs to be strengthened, but not at the cost of the domestic producers. Food-processing sector should be directed to use the abundant raw material available within the country. The 'rainbow' revolution that everyone talks about is actually aimed at helping the industry to exploit the farm sector. Already a number of manufacturing units, for instance, have begun to source the agricultural raw material, including oranges, grapes, popcorn, peas etc, from America and Europe. Domestic production in these crops is going waste. Farmers have repeatedly and in different parts of the country been dumping tomatoes, potatoes and other fruits onto the streets to express their frustration at the lack of adequate marketing infrastructure.

Creating a global market for farm produce is the bane of modern agriculture. The seed multinationals, the food giants, and the supermarkets, have cornered the food chain in the process thereby destroying livelihoods, local markets and also drastically reducing food choices. Such a maket strategy has resulted in the disappearance of locally produced nutritious foods as a consequence of which micro-nutrient deficiency in human populations have grown manifold. Encouraging local markets will also reduce the dependence upon long distance transportation thereby minimising global warming. It will also help in bringing back the traditional and neglected crops, and help in changing the food habits.

Farm incomes: Growing indebtedness in agriculture is forcing an increasing numbers of farmers to end their lives. This unsavory phenomenon is a manifestation of the declining farm incomes and lack of farm credit. Institutional finance and credit has almost disappeared over the years. Banks are no longer treating agriculture for priority sector lending. Rural Banks and cooperatives are deep in the red, with a majority of them eating into their own reserves. Agriculture credit has to be revived.

Schemes that encourage banks to provide easy credit facilities to farmers need to be spelled out. On top of it, agriculture credit has to be extended to sustainable farming systems. So far the banks are only providing credit for technology-oriented farming systems. This has to be extended to organic agriculture, for which an Organic Bank need to be created by NABARD (like the technology credit that goes through the private Robo Bank). Crop insurance should be extended to cover the entire farm sector immediately.

Food Security – The Way Ahead

Although, India is following the WTO dictates of doing away with the food procurement system, any tinkering with what is generally regarded as the "famine-avoidance" strategy, can be catastrophic. Corrective measures are needed to reduce inefficiency in the system while at the same time making it broad-based and widespread.

Multiple Cropping: Emphasis on commodities approach during the green revolution has encouraged monocultures, loss of biodiversity, encouraged food trade in some commodites, distorted domestic markets, and disrupted the micro-nutrient availability in soil, plant, animals and for humans humans. Thrust on farm commodities have also pushed in trade activities, encouraged food miles, adding to greenhouse emissions, water mining, and destruction of farm incomes. The need is to revert back to the time-tested farming systems that relied on mixed cropping and its integration with farm animals, thereby meeting the household and community nutrition needs from the available farm holdings.

Reverting back to mmultiple cropping will also provide the answer to the acute malnutrition that prevails in the countryside. The availability of nutritious crops, vegetables and fruits was once a part of the cropping pattern, abandoned in the wake of green revolution. The second green revolution that is being talked about will further exacerbate malnutrition crisis. This can only happen when the focus shifts away from encouraging cash crops.

For the past two decade at least, the World Bank/IMF and some other academicia and donors have been pressing developing countries to diversify from staple foods to cash crops in what is being projected as the right approach to added to farm incomes. This is a politically motivated advise and runs counter to the sustainable approached spelled out above. Many Latin American countries are faced with a serious land degradation crisis as a result. It also pushes farmers into a death trap since the developing countries do not have the resources to provide for adequate marketing infrastructure.

Public Distribution System (PDS) also needs to be strengthened and extended to upcoming agricultural areas in Bihar, Orissa, West Bengal and the northeast. Similarly, financial allocation must be made for assured food procurement at remunerative prices. In addition, procurement needs to be extended to coarse cereals, pulses and oilseeds to provide farmers an incentive to produce more. Food procurement operations, linked to the announcement of assured prices for agricultural commodities, were the two planks of the 'famine-avoidance' strategy that India had adopted in the wake of the green revolution. Whether the economists like it or not, the fact remains that a combination of these policies helped India to emerge from the dark days of 'ship-to-mouth' existence.

The emphasis by IMF, the World Bank and WTO to force India dismantle PDS is based on the corporate need. India's massive food procurement operations are coming in the way of the expansion of the food trade that the United States and the European Union are looking for. If the US doesn't find an assured food market in a country as huge as India, with one sixth of the world's population, the chances are that its own agriculture will collapse under the artificial weight of its own federal subsidies.

That the threat is real, is clearly evident. Take a look at the recent developments in neighbouring Pakistan. Under pressure from IMF and the World Bank, Pakistan's military government has begun lifting its decades-long support price system for key commodities – despite protests that this would be disastrous for small farmers. In India too, economists are asking the government to 'decentralise' the food procurement system, a euphemism for dismantling the PDS.

Once the government withdraws from announcing procurement prices for agricultural commodities, it is under no obligation to purchase the surplus that flows into the *mandis*. Farmers would thus be left at the mercy of the trade and the market forces, and if the past experience is any indication it simply means rendering the farming community vulnerable to exploitation thereby threatening the country's food self-sufficiency, so assiduously built over the past three decades.

The biggest crisis afflicting the farm sector is the inability to manage the agricultural surpluses. It is here that the policy planning effort has to be redirected with an effort to ensure that the surplus does not become a national liability. The approach has to be different for the rural and urban areas. Since this paper focuses on the link between agriculture, food security and hunger, a framework for rural India is hereby proposed.

Community Grain Banks: The answer to the intricately complex, economically unsound and politically sensitive issue of public distribution rests with the poorest of the poor and is a tribute to human ingenuity, cooperation and traditional knowledge. Effectively targetting the public distribution system to reach the needy and the poorest of the poor has been a serious concern. Moreover, for several years now, the exclusion of the well-to-do beneficiaries, including income tax payers, from the provisions of the PDS have been resisted by all political parties, irrespective of their ideological leanings.

While the debate goes on, Bolangir in Orissa and Kodagu in Karnataka have demonstrated that the real beneficiaries, the poor in the villages, are not dependent upon food doles. Such a system of sharing the benefits of the harvest with the village community also exist in several other parts of the country. This is perhaps the only viable path for the nation to wriggle out of the growing threat from food insecurity.

Starvation and hunger no longer stalks a cluster of 20 villages, about 150 kms away from Bolangir town. At a time when recurring drought has brought acute misery and suffering for tens of thousand people in the district, and with the latest controversy shrouding the starvation deaths and sale of children from western Orissa showing no signs of healing, hundreds of families in and around Sundhi munda village have built a food insurance system that keeps sure hunger and death at bay. That the food security system has successfully withstood varying degrees of natural calamities and has, in fact, grown and multiplied clearly demonstrates its social relevance and effectiveness.

It all began in 1990-91, when a social activist Bansi Dhar Behera, coordinator of the *Anchalika Jana Sewa Anusthan* in Sundhi munda village, was looking for a permanent solution to mitigate human suffering arising from the non-availability of foodgrains, especially at times of distress. His appeal to fellow villagers to donate surplus paddy and rice after the harvest so as to build a grain reserve brought in 22 quintals of paddy. In all, 150 families from eight villages, almost all of them marginal farmers, responded to his call. The village grain bank was thus formed.

The grain bank became a pivot of food security. Farmers have since then deposited their 'surplus' produce with the bank after each paddy harvest. They withdraw an equal quantity of paddy at the time of need without having to pay any interest. For others, who are landless or do not have any 'surplus' for the grain bank, borrowing paddy at the time of distress is a routine. But at the time of harvest, the grains borrowed have to be returned with half a bucket of paddy as interest. For those, who cannot repay the foodgrain loan, the village *samaj* decides whether the loan can be waived or not. For the villagers, the grain bank was an escape from the clutches of the money-lenders, who often gave foodgrains to the needy to be returned in double the quantity received, and that too within three months.

Sometimes, depending upon the immediate requirement of the participating villages, the beneficiaries are asked to contribute by way of human labour. In village Batharla, a community temple and a grain store house was constructed by the beneficiaries. Their wages were paid in kind from the interest (surplus grain) that builds up over the years. In Banjupadhar village, a traditional water harvesting tank was rejuvenated for which the society distributed 16 quintals of paddy as wages. The grain bank, in other words, is also being utilised for 'food for work' programmes, all depending upon the need of the village community.

In five years, the grain bank had grown in size and volume. In 1996, the society received and disbursed 220 quintals of paddy. A year later, in 1997, it got back 253 quintals. In all, the number of people donating to the grain bank had grown by almost ten times, with a thousand families depositing paddy this year. The number of beneficiaries too increased over the years reaching 1,066 families this year, in the 20 participating villages. More than the numbers what is important is to understand that these families have perfected a social model that gives them the freedom from hunger.

The ten grain banks in Kodagu district are, however, registered under the Cooperative Act. Successfully in operation for over 30 years now, these grain banks also work on the same principle. After every paddy harvest, each member brings not less than 100 kg of paddy as their contribution to the grain bank. And during the lean months of December-January, paddy can be borrowed as loan by members. The loan is normally repaid after the next harvest with an interest of 12 per cent in terms of paddy. After the harvesting season ends, the left over paddy stocks are sold in the market. Consequently, members receive dividend varying between 10 to 20 per cent of the total share capital.

Such is the underlying spirit of cooperation that like in Bolangir, each member in Kodangu district also deposits about five to ten kg of paddy every year towards what is called as the 'death fund'. The basic idea being that at times of bereavement, the village community comes to the rescue of the family in mourning. It is invariably because of the strong community ties in the villages that the grains banks have succeeded. Also, because these grain banks have remained outside the gambit of government interference. Its replication, therefore, has to be through the panchayats and the grassroot NGOs or perhaps an amalgamation of both.

Village Republics: Focus on tackling the causes of poverty, hunger, the inequitable distribution of income and low human resource base with the objective of providing everyone with the opportunity to earn a sustainable livelihood. The green revolution areas are encountering serious bottlenecks to growth and productivity. Excessive mining of soil nutrients and groundwater have already brought in soil sickness. If the livelihood of the marginalised in the society (and that in the majority world is in agriculture) it must be secured by economic activities that are sustainable, that do not threaten the integrity of the environmental assets on which they depend. Food security and hunger are directly linked to the community's control over the natural resources, and also on the long-term sustainability of the resource base.

Contrary to commonly made projections and assessments, hundreds of villages in rural India have made their own effort to chart a different but equitable path to growth and human development. Deviating from the mainstream approach, these villages have put up sign board outside the village boundary warning government officials and private company executives from entering their village. The reason: these villages have become self-reliant.

A conservative estimate based on different reports shows that close to 1500 villages have imposed self-rule and have declared themselves village republics. In these villages the residents have taken control over their natural resources – namely forest, land, minerals and water sources – and have formed strong institutions to manage them. They plan, execute and resolve all affairs inside the village and government officials and programmes are accepted only after getting approval of the residents through *Gram Sabha* (village assembly consisting of all adult members). In many such villages, the forest department, the police and other officials just execute programmes and plans chalked out in village meetings.

Self-reliant villages is the answer to India's multiple and complex problems of food insecurity, hunger and malnutrition.

10

THE NARMADA VALLEY

INTRODUCTION

The Narmada is the longest west flowing river in the country, traversing over 1300 kms through three states- M.P., Maharashtra and Gujarat before it meets the sea. The discharge of the Narmada is said to be more than the combined discharge of the Sutlej and the Beas. It is certainly the most important river valley for the central and western India.

The plans to "develop" the river have been in existence since many years. The first concrete plans for the series of large dams now planned on the river emerged sometime around 1946. Since then, these plans have gone through a series of interstate disputes, changes and transformation. Today, the grandiose Narmada Valley Development Project (NVDP) consists of 30 big dams, 135 medium dams and over 3000 "small" dams on the river and its tributaries. These include the two gigantic projects Sardar Sarovar, (SSP) in Gujarat and the Narmada Sagar (Indira Sagar) in M.P. Except for the SSP all the other project are in Madhya Pradesh.

The NVDP has been in existence for at least a few decades, and certainly since 1979 when the Narmada Water Disputes Tribunal (NWDT) gave its Award, fixing the heights of the SSP and NSP and the shares of the Narmada waters of the party states. However, even 16 years after the Tribunal Award, the implementation of the NVDP and the two mega projects in particular, is woefully slow.

Only two of the big dams, Tawa and Bargi have been completed, and a few medium ones like Sukta, Barna. Some medium ones like Man, Jobat are nearing completion. However, even in cases where dams have been completed, the projects have not. In Bargi, for example, the dam has been completed, but the canals are ready for only a negligible area. The SSP dam work has been suspended since year and a half. The work on the NSP, Maheshwar and Omkareshwar dams - the kingpin of the NVDP - has never really taken off mainly due to the inability to resolve with the two issues - of severe financial crunch and massive displacement. The international funding agencies too are keeping away from the Narmada.

Today, there are now serious doubts about even the very feasibility of the completion of the NVDP. In fact, the state of Gujarat is basing its plans for the SSP on this - that Madhya Pradesh is never going to be able to use its share of water, which then would accrue to Gujarat.

It is not too difficult to see the reasons for this state of affairs.

The Development Paradigm of NVDP

The NVDP is clearly based on a development paradigm that was dominant during the time it evolved - i.e. the 40's and the 50's. This development paradigm has some very characteristic features, many of which are directly responsible for the state of affairs today. These include:

- Water to be collected in centralized storage and then distributed through extensive canals
- Large scale/ mega projects with massive storage
- Oriented towards commercial energy
- Whole outlook was how to use all the water that is "going waste"
- Enormous social and environmental impacts, many which society was supposed to absorb ("some one has to sacrifice for the development", "you can't make an omelet without breaking an egg" principle)

- Massive and ever- escalating financial outlays
- Uncertainty, over-estimation and concentration of benefits

The Impacts, and Need for an Alternative

A direct result of all the above has been the increasing financial unviability of projects, the severe time and cost overruns, floundering projects, projects that can't take off and projects that get stuck half way because of financial problems.

On the other hand has been the trail of destruction and human misery that the projects have left behind, as they have displaced lacs of people, disintegrating families, hamlets, villages and whole communities, destroying well established economic support systems and even whole cultures. Mounting antagonism against these large dams has been the result, as those ousted and those to be ousted are no longer willing to take the injustice lying down. And so we have the emergence of strong resistance movements, which are bound only to grow, not diminish. A new phenomenon is the emergence of organised protest and struggle for justice that is taking place around projects that have been completed long back. The ongoing struggles of the oustees of Tawa and Bargi dams, displaced between two to half a decade back are a case in point showing clearly that nowhere has just and proper rehabilitaltion or resettlement been achieved. This resistance is bound to get stronger and more bitter, more aggressive, as more and more projects, involving more and more displacement are sought to be brought on line.

The rising consciousness among the people, the Government and other institutions about the environmental impacts of such projects has also made it difficult for these projects to get clearances from the environmental angle.

Further, serious doubts have been raised about the very rationale of these projects - i.e. - their benefits themselves. Accumulating volumes of evidence show that the benefits of these projects always tend to be overestimated, and the cost, including financial, under-estimated. The project benefits are almost invariably distributed in an inequitable manner, so much so that these projects (like the SSP) have been described as "State funded creation of inequity".

Large irrigation systems are resultting in waterlogging and salinisaiton of the very lands that they were supposed to benefit. Further, it has pulled the farmers firmly in to the orbit of "chemical agriculture", whcih is slowly but surely destroying ther fertility of the land, and on the other side, pushing the farmers heavily in to debt.

The whole NVDP seems to have stagnated, and there is little chance of it being able to move ahead, if today's plans persist, given the fact that all the above mentioned factors are likely to accentuate.

Thus, to be able achieve any realistic progress in the Narmada Valley, there is an urgent need to formulate alternative plans, plans that will avoid the above constraints, and will thus allow for a speedy, sustainable and equitable development of the Narmada Valley.

Such an *Alternative Framework Of Development For The Narmada Valley* is possible. However, it would mean a significant departure from the type of plans that constitute the NVDP, though many - even most - of the ideas, technologies for this are nothing new and have been a part of what even the Government has been advocating - albeit - in a piecemeal manner, over the past decade or so at least.

Principles/ Goals of the New Paradigm

The goals, or the principles on which the alternative model of development of the Valley would have to be based include:

- Sustainability, and Eco-enhancement
- Minimum or Negligible Displacement; No Uprootment
- Focus on the Prioritization of Use of Water (Drinking and Domestic Use, Cattle, Protective Irrigation, Commercial Crop/Local Industry, Industry -and Long Distance Transport, Ecological Functions)
- Equity

- Integrated Land-Water-Energy Management; and a "Farming systems" approach, rather than concentrating on one or two crops
- Decentralized Decision making structures, and Control in Hand of Local Community

Evolving the New Paradigm

The major elements of the Alternative Plan based on the above goals are by now well known. *We strongly argue that comprehensive integrated watershed management and in-situ rain water harvesting, on an extensive scale all over the valley, combined with groundwater development and small, and micro irrigation scheme can form the backbone of a just and sustainable development plan for the Narmada Valley.*

This would include the following elements:

- Integrated Watershed Management
- In-Situ Rain Water Harvesting Schemes, Soil Water Conservation
- Large Scale Eco-Enhancement And Raising Of Carrying Capacity
- Small And Micro Storage, Lift Irrigation Schemes
- Groundwater Development And Utilization
- Run Of The River Schemes On Tributaries
- Decentralized Options For Energy Generation Including Biomass Based, Solar, Micro-Hydel Etc.

Really speaking, there is no longer anything "new" in this paradigm. Even the Government now states all these principles as the basis of their development plans. Yet, it also persists with large scale mega projects as the central pillar of its plans. The need is to resolve this contradiction.

The key shift is to see these not as just small add-ons, secondary to the large projects, but to see that a combination of all the above implemented on an extensive scale can form a comprehensive development plan for the Valley, achieving far better results at a far lesser cost. It is thus a matter of giving these technologies their legitimate role. Moreover, these technologies would have to be comprehensively integrated with conservation and efficiency measures like the following:

- Conservation and Control of Wasteful Use of Water/ Energy
- More efficient Use Water/ Energy
- Priority to the Utilization of the irrigation Potential already created, for which large parts of the financial, social and environmental costs have already been paid.
- Severe Control on Pollution of Water resources
- Industries to recycle and reuse their water to the full extent possible

  Further, for the time being, some elements of conventional technologies which are relatively benign may also be incorporated in the plan, on a transitional basis. These may include:

- High head pumped storage schemes
- Gas based power stations

Emerging Consensus

A consensus is slowly but surely building up that such an approach is necessary, is possible, and in fact, inevitable.

For example, the Planning commission of India initiated in 1988 an exercise called The Agro Climatic Regional Planning. This exercise divided the Country [Planning commission, 1989] "into 15 Regions delineated on the basis of commonality of agro-climatic factors like soil type, rainfall, temperature, water resources, etc....In this innovative approach based on agro-climatic zones, an overall development profile of each region is formulated through an optimal mix of land stock management, crop production, animal husbandry, aquaculture, horticulture, forestry and agro-processing industry."

The Planning Commission exercise also made it clear that "A farming system approach at the disaggregated level, rather than the traditional crop production approach is being attempted."

The Narmada Valley falls into four of these 15 zones.

Zone VII: Eastern Plateau and Hill Region (Districts Shahdol, Balaghat, Durg)

Zone VIII: Central Plateau and Hills Region (Including Districts Mandla, Jabalpur, Narsingpur, Raisen, Hoshangabad, Chindwara, Sagar, Sehore, Betul)

Zone IX: Western Plateau and Hill Region (Districts Dhar, Khandwa, Jhabua, Khargone (all M.P.) and Dhule, Maharashtra)

Zone XIII: Gujarat Plains and Hills Region (Vadodara, Bharuch)

The report then goes on to describe in detail the development strategies for these regions.

For Zone VIII, the Report says (Page 79):

"3.98: The undulating topography, underdeveloped irrigation potential, and a large proportion of area under rainfed farming suggest water conservation, crop diversification, ground water development and input supplies services to be the thrust areas for proper development and growth of this zone".

Similarly, for Zone XIII (Pg. 89):

"In view of nearly 76 % being rainfed farming, the major thrust should be on rainwater harvesting and management, dryland farming, canal and groundwater management".

Further on, the report emphasizes in detail the role, importance and need for integrated watershed management, minor irrigation including groundwater development, crop planning as the key thrust areas.

The Planning Commission report is presented as it represents an exercise by the highest planning body in the Country. However, numerous authors, papers and field workers are now converging towards a similar point of view.

Issues for Discussion

While many of the above ideas are increasingly gaining all round acceptance, they still need to be integrated and formulated into comprehensive plans covering the large areas like the Narmada Valley. In terms of evolving the specific plans, the following issues will have to be discussed:

- Steps in Evolving Detailed plans for the Valley, and the sub-regions within it
- Preliminary Quantification, Outline Plans
- Criteria for Evaluation of the Alternative Plan
- How to make this kind of planning into a people's initiative and people's movement

11

AGRICULTURAL DEVELOPMENT IN INDIA

An appreciable pace of research has taken place in all sectors of agriculture including crops, horticulture, natural resource management, livestock, fisheries and agricultural engineering. The technology-led developments in agriculture have made India self-sufficient in foodgrains and a leading producer of several commodities in the world. The green revolution in crops, yellow revolution in oilseeds, white revolution in milk production, blue revolution in fish production and a golden revolution in horticulture bear an ample testimony to the contributions of agricultural research and development efforts undertaken in the country.

India has received worldwide acclaim in the field of agricultural research and education. The National Agricultural Research System (NARS), with the Indian Council of Agricultural Research (ICAR) as an apex body, has been striving for a holistic development of agriculture at the national level by planning, promoting, conducting and co-ordinating research, extension and education on all aspects of agriculture including animal sciences and fisheries for ensuring the optimal utilization of land, water, plant and animal genetic resources for sustainable agricultural development.

The Council undertakes research on fundamental and applied aspects in traditional and frontier areas to offer solutions to problems relating to conservation and management of resources, productivity improvement and health management of crops, animals and fisheries, post-harvest technology and value addition. This helps in developing new technologies in agriculture and allied sectors. The Council with its headquarters at New Delhi has a vast network of institutes all over the country consisting of 49 institutes including 5 National Institutes with 4 having Deemed University status, 5 National Bureaux, 33 National Research Centres, 11 Project Directorates and 78 All-India Co-ordinated Research Projects. Besides, it coordinates agricultural education in the country through a network of 34 State Agricultural Universities, 1 Central Agricultural University, 3 Academic Universities and 5 Deemed Universities.

All activities under the transfer of technology programme of the Council have been integrated with Krishi Vigyan Kendras (KVKs) and the mandate enlarged to perform the function of on-farm testing and research, long-term vocational training, in-service training of grass-root level functionaries and frontline demonstrations. A network of 344 KVKs has so far been established in our country. A strong back-up support is provided through 8 Trainers' Training Centres to these KVKs by updating knowledge of the KVK trainers in the advancements made in research.

To strengthen research extension efforts, implementation of Institution-Village Linkage Programme (IVLP) at 70 centres in the country, establishment of 40 Agricultural Technology Information Centres (ATIC), strengthening of 53 Zonal Agricultural Research Stations (ZARS) to take up the additional functions of KVK, and Strengthening of Directorate of Extension of State Agricultural Universities and Zonal Units have been recently undertaken under the National Agricultural Technology Project (NATP).

The Council aids, promotes and coordinates agricultural education programmes at the national level. The educational programmes are carried out through its network of universities. In addition to these, 4 National Institutes of the Council, viz., Indian Agricultural Research Institute (IARI), Indian Veterinary Research Institute (IVRI), National Dairy Research Institute (NDRI), and Central Institute of Fisheries Education (CIFE) have been granted the status of deemed university. These institutes offer post-graduate degree programmes besides undertaking research in their respective areas.

Achievements

The agriculture research system has significantly contributed to productivity growth in almost all the sectors of agriculture. For instance with reference to 1950, the productivity gains are nearly 3.3 times in foodgrains, 1.6 times in fruits, 2.1 times in vegetables, 5.6 times in fish (aquaculture), 1.8 times in milk and 6.4 times in eggs. Development of about 3200 high yielding improved varieties of different crops and their production technologies and preservation of 2.2 lakh germplasm accessions of agri-horticultural crops and their wild relatives for future use in breeding programmes, are some remarkable achievements of the National Agricultural Research System. It is also revealing to note that the modern technologies have contributed to saving of the area, other resources and in increasing revenues and exports. For instance, modern varieties of rice saved nearly 39 million hectare and the wheat varieties 37 million hectare area.

Research in livestock and poultry sector has provided more options for increased income generation. The fisheries sector is also growing at an incremental pace, with an annual production of 5.9 million tonnes. Along with the livestock and poultry sector, it also contributes significantly to domestic nutritional security and rural development.

Research for development of farming implements has helped to make agricultural operations modern and less labour-intensive apart from saving the operational cost.

The programme on frontline demonstrations of improved production technologies, on-farm demonstrations of integrated nutrient and pest management, and a variety of training programmes involving farmers have shown encouraging results by successfully demonstrating the potential of new and improved technologies.

Keeping India Vision 2020 in view, each ICAR institution has formulated its Perspective Plan document, 2020. Adopting a bottom up approach, ICAR has also formulated its Vision 2020 document which has laid a broad road map for the next 20 years. ICAR scientists and peers have strived hard to set up a demand-driven research and development agenda. The gains from green revolution were not only consolidated but also several new grounds were broken for attaining new gains. Technological improvements were made in several non-traditional crops like soyabean, new crop seasons were explored in rabi maize, rabi sorghum, summer groundnut and summer pearl millet. Hybrid breeding approach was extended to rice, safflower, castor, pigeon pea and mustard.

Increased yields and stability were achieved through shifting away from kharif season production to rabi season production. Diversification covering fruits and vegetables, flowers, animal enterprises, poultry, fish and agro-forestry are also emphasised aiming at golden revolution. All these efforts added to defending the gains of green revolution, extending the gains to new areas and making new gains in several important areas contributing to accelerated agricultural transformation.

GROWING DEMAND

Notwithstanding the phenomenal success in agriculture, increasing population and per capita income demand higher production and productivity levels. The annual increase in demand by the end of the 10th Five Year Plan will be 2.6 million tonnes in rice, 2.2 million tonnes in wheat, 1.6 million tonnes in pulses, 4.2 million tonnes in fruits, 2.5 million tonnes in vegetables, 7.8 million tonnes in milk and 0.6 million tones in fish. The 10th Plan fixes a target growth rate of 4 per cent in agriculture which in itself is challenging, keeping in view the achieved growth rate of 1.95 per cent in food production during the IX Plan.

After attaining self-sufficiency in food, concerns on increasing factor productivity, improving resource efficiency and economic access to food combined with better nutrition, health and shelter became important which were addressed through diversification of agriculture. At this point of time, natural resource management problems became critical, leading to plateauing of yield and, therefore, conservation of natural resources such as soil, water and genetic wealth. The concerns of global competitiveness and system's sustainability became critical around 2000 which are now receiving priority attention.

The eight-fold current concerns of the ICAR include –input use efficiency and consequent reduction in cost of cultivation; diversification of agriculture while conserving natural resource base; rainfed agriculture; post-harvest management; sustainability, equity and gender issues; motivated and creative human resource; easy access and quick disposal of information and building a culture of trust and sense of belonging among scientists, farmers, extension workers, young entrepreneurs and NGOs. Thus, the concerns are evolving and the system is responding well. The contributions of technology in particular have been unique.

TECHNOLOGY LED-PROSPECTIVE AGRICULTURAL GROWTH

Indian agriculture is passing through many challenges following increasing demands of foodgrains, growing competition among nations for export of value-added quality produce, rising cost of production and dwindling sources of water. But they offer new opportunities for quality improvement, value-added products development, processing and enhancing the shelf life of agri-products through which India can compete with the rest of the world. New technologies like biotechnology, hybrid technology and bio-informatics will have their imprint on our success for meeting the challenges of food security and sustainability.

The six new initiatives of the Department of Agricultural Research and Education during the Xth Plan, are – Network Project on organic farming, Impact of Climate Change on Agricultural Productivity, Networking on Transgenics, National Bureau of Indian Veterinary Type Culture, Network on Biosystematics of Insects and National Research Centre for Agricultural Extension. Rainbow revolution encompassing an all-round growth in the production of foodgrains, edible oils, fruits, vegetables and animal and fish products and evergreen revolution in agriculture through diversified agriculture, precision farming, resource conservation and value addition to provide in real sense, the required food, nutrition and environmental security are the hopes of the future.

PRIORITIES FOR AGRICULTURE AND RURAL DEVELOPMENT

Although agriculture contributes only 21% of India's GDP, its importance in the country's economic, social, and political fabric goes well beyond this indicator. The rural areas are still home to some 72 percent of the India's 1.1 billion people, a large number of whom are poor. Most of the rural poor depend on rain-fed agriculture and fragile forests for their livelihoods.

The sharp rise in foodgrain production during India's Green Revolution of the 1970s enabled the country to achieve self-sufficiency in foodgrains and stave off the threat of famine. Agricultural intensification in the 1970s to 1980s saw an increased demand for rural labor that raised rural wages and, together with declining food prices, reduced rural poverty.

Sustained, although much slower, agricultural growth in the 1990s reduced rural poverty to 26.3 percent by 1999/00. Since then, however, the slowdown in agricultural growth has become a major cause for concern. India's rice yields are one-third of China's and about half of those in Vietnam and Indonesia. With the exception of sugarcane, potato and tea, the same is true for most other agricultural commodities.

The Government of India places high priority on reducing poverty by raising agricultural productivity. However, bold action from policymakers will be required to shift away from the existing subsidy-based regime that is no longer sustainable, to build a solid foundation for a highly productive, internationally competitive, and diversified agricultural sector.

ISSUES AND CHALLENGES

Slow Down in Agricultural and Rural Non-Farm Growth: Both the poorest as well as the more prosperous 'Green Revolution' states of Punjab, Haryana and Uttar Pradesh have recently witnessed a slow-down in agricultural growth. Some of the factors hampering the revival of growth are:

- Poor composition of public expenditures: Public spending on agricultural subsidies is crowding out productivity-enhancing investments such as agricultural research and extension, as well as investments in rural infrastructure, and the health and education of the rural people. In 1999/2000, agricultural subsidies amounted to 3 percent of GDP and were over 7 times the public investments in the sector.
- Over-regulation of domestic agricultural trade: While economic and trade reforms in the 1990s helped to improve the incentive framework, over-regulation of domestic trade has increased costs, price risks and uncertainty, undermining the sector's competitiveness.
- Government interventions in labor, land, and credit markets: More rapid growth of the rural non-farm sector is constrained by government interventions in factor markets — labor, land, and credit — and in output markets, such as the small-scale reservation of enterprises.
- Inadequate infrastructure and services in rural areas.

Weak Framework for Sustainable Water Management and Irrigation:

- Inequitable allocation of water: Many states lack the incentives, policy, regulatory, and institutional framework for the efficient, sustainable, and equitable allocation of water.
- Deteriorating irrigation infrastructure: Public spending in irrigation is spread over many uncompleted projects. In addition, existing infrastructure has rapidly deteriorated as operations and maintenance is given lower priority.

Inadequate Access to Land and Finance:

- Stringent land regulations discourage rural investments: While land distribution has become less skewed, land policy and regulations to increase security of tenure (including restrictions or bans on renting land or converting it to other uses) have had the unintended effect of reducing access by the landless and discouraging rural investments.

- Computerization of land records has brought to light institutional weaknesses: State government initiatives to computerize land records have reduced transaction costs and increased transparency, but also brought to light institutional weaknesses.
- Rural poor have little access to credit: While India has a wide network of rural finance institutions, many of the rural poor remain excluded, due to inefficiencies in the formal finance institutions, the weak regulatory framework, high transaction costs, and risks associated with lending to agriculture.

Weak Natural Resources Management: One quarter of India's population depends on forests for at least part of their livelihoods.

- A purely conservation approach to forests is ineffective: Experience in India shows that a purely conservation approach to natural resources management does not work effectively and does little to reduce poverty.
- Weak resource rights for forest communities: The forest sector is also faced with weak resource rights and economic incentives for communities, an inefficient legal framework and participatory management, and poor access to markets.

Weak delivery of basic services in rural areas:

- Low bureaucratic accountability and inefficient use of public funds: Despite large expenditures in rural development, a highly centralized bureaucracy with low accountability and inefficient use of public funds limit their impact on poverty.

In 1992, India amended its Constitution to create three tiers of democratically elected rural local governments bringing governance down to the villages. However, the transfer of authority, funds, and functionaries to these local bodies is progressing slowly, in part due to political vested interests. The poor are not empowered to contribute to shaping public programs or to hold local governments accountable.

PRIORITY AREAS FOR THE WORLD BANK SUPPORT

Enhancing Agricultural Productivity, Competitiveness, and Rural Growth

Enhancing productivity: Creating a more productive, internationally competitive and diversified agricultural sector would require a shift in public expenditures away from subsidies towards productivity enhancing investments.

Second it will require removing the restrictions on domestic private trade to improve the investment climate and meet expanding market opportunities.

Third, the agricultural research and extension systems need to be strengthened to improve access to productivity enhancing technologies. The diverse conditions across India suggests the importance of regionally differentiated strategies, with a strong focus on the lagging states.

Improving Water Resource and Irrigation/Drainage Management: Increase in multi-sectoral competition for water highlights the need to formulate water policies and unbundle water resources management from irrigation service delivery. Other key priorities include:

(i) modernizing Irrigation and Drainage Departments to integrate the participation of farmers and other agencies in irrigation management;

(ii) improving cost recovery;

(iii) rationalizing public expenditures, with priority to completing schemes with the highest returns; and

(iv) allocating sufficient resources for operations and maintenance for the sustainability of investments.

Strengthening rural non-farm sector growth: Rising incomes are fueling demand for higher-value fresh and processed agricultural products in domestic markets and globally, which open new opportunities for agricultural diversification to higher value products (e.g. horticulture, livestock), agro-processing and related services.

The government needs to shift its role from direct intervention and overregulation to creating the enabling environment for private sector participation and competition for agribusiness and more broadly, the rural non-farm sector growth.

Improving the rural investment climate includes removing trade controls, rationalizing labor regulations and the tax regime (i.e. adoption of the value added tax system), and improving access to credit and key infrastructure (e.g. roads, electricity, ports, markets).

Improving Access to Assets and Sustainable Natural Resource Use

Balancing poverty reduction and conservation priorities: Finding win-win combinations for conservation and poverty reduction will be critical to sustainable natural resource management. This will involve addressing legal, policy and institutional constraints to devolving resource rights, and transferring responsibilities to local communities.

Improving access to land: States can build on the growing consensus to reform land policy, particularly land tenancy policy and land administration system. States that do not have tenancy restrictions can provide useful lessons in this regard. Over the longer term, a more holistic approach to land administration policies, regulations and institutions is necessary to ensure tenure security, reduce costs, and ensure fairness and sustainability of the system.

Improving access to rural finance: It would require improving the performance of regional rural banks and rural credit cooperatives by enhancing regulatory oversight, removing government control and ownership, and strengthening the legal framework for loan recovery and the use of land as collateral. It would also involve creating an enabling environment for the development of micro-finance institutions in rural areas.

Strengthening Institutions for the Poor and Promoting Rural Livelihood

Promoting Community-Based Rural Development: State Government efforts in scaling up livelihood and community-driven development approaches will be critical to build social capital in the poorest areas as well as to expand savings mobilization, promote productive investments, income generating opportunities and sustainable natural resource management.

Direct support to self-help groups, village committees, user's associations, savings and loans groups and others can provide the initial 'push' to move organizations to higher level and access to new economic opportunities. Moreover, social mobilization and particularly the empowerment of women's groups, through increased capacity for collective action will provide communities with greater "voice" and bargaining power in dealing with the private sector, markets and financial services.

Strengthening Accountability for Service Delivery: As decentralization efforts are pursued and local governments are given more prominence in the basic service delivery, the establishment of accountability mechanisms becomes critical. Local governments' capacity to identify local priorities through participatory budgeting and planning needs to be strengthened. This, in turn, would improve the rural investment climate, facilitating the involvement of the private sector, creating employment opportunities and linkages between farm and non-farm sectors.

GLOBALIZATION AND INDIAN AGRICULTURE: FROM HOPE TO DESPAIR?

In my treatment of the vast theme of globalization, I will limit myself to the agricultural sector of my country, India. My thesis is simple. And although no one should believe a person who begins by claims to simplicity, I argue that until recently, globalization in Indian agriculture took the form of international cooperation.

Further, this collaboration constructed the base for the recent entry of private, for-profit foreign investment. I also argue that this earlier globalization was indeed effective, together of course with steps taken within the country, in "cultivating peace" that is, in a country with many reasons for violence, agricultural prosperity indeed averted large-scale conflicts.

In contrast to that period of internationality, the current globalization risks much if it does not include a strong, sustainable plan for those who are the most vulnerable in the processes of development.

The Earlier "Globalization". International collaboration in Indian agriculture had already begun in 1957 when the Rockefeller Foundation and the Indian Government began a program to improve maize.

Because of this early link with foreign collaborators, the country received varieties of wheat and rice from, respectively, Mexico (www.cimmyt.org) and the Philippines, (www.irri.org) in the early Sixties.

Five years later in India, the Green Revolution was launched (which, alas, is not at all related to the Green of the Esperanto flag, I must quickly clarify! The technological package included better, higher-yielding crop varieties; use of chemical fertilizers; modern irrigation systems; and strengthened programs which supplied farmers with practical knowledge.

Also at that time was established in the southern Indian city of Hyderabad the institute where I now work, ICRISAT (www.icrisat.org) — the English acronym for an internattional research institute for agriculture in the semi-arid tropics. This institute, with the 16 others which are members of the Consultative Group for International Agricultural Research have contributed much to agricultural growth not only in India but also in the rest of the developing world.

For example, in the 25 years of its existence, ICRISAT's partnership with national agricultural systems developed around 370 varieties of grains and legumes for distribution in more than 70 countries; collected and conserved from 130 countries, the seeds of more than 113 *thousand* varieties of these six grains and legumes; and ICRISAT has trained more than 3000 scientists from 90 countries.

A Strong Infrastructure. This kind of successful collaboration presupposes the existence of a sufficiently strong infrastructure for agriculture, which India indeed possesses. Of course, in a country where about 65% of the population finds a livelihood in agriculture and related sectors, any general raising of standard of living must necessarily involve the rural sector.

The so-called "Five-Year Plans" of the economy (from 1951 onwards) have, in fact, done exactly that. Much investment both human and financial was made in agriculture. The great part of this investment was capital investment, for example, roads, and large irrigation projects.

This large-scale construction of agriculture infrastructure too received much support from the "first world". For instance, in the 25 years between 1962 and 1987, just the irrigation sector received 3800 million dollars from the World Bank.

India also received help in the design and implementation of policies which aimed to protect the most vulnerable sections of the people.

The main policies included: price-support for farmers; establishing "fair-price shops"; policies regarding the procurement of agricultural produce; controlling the flow of food across state boundaries; and a flexible import policy that allowed large imports when it became necessary.

This combination of infrastructure, science, and policy indeed averted both catastrophic famines of the earlier kind, and large-scale unrest. And reasons for unrest there were (and indeed are!) aplenty: India is home to 15% of the world's population; and to a quarter of the population of developing countries. As already mentioned, 65% Indians live in the rural sector where poverty is widespread: 53% of the inhabitants have less than 1 dollar a day to spend.

Unjust Land Distribution. The rural violence which did occur during the last 50 years of Independent India of course has complex roots in history and culture, but that violence also shows how easy it is for violent conflict to occur in underdeveloped regions.

The "decade of chaos" from 1967 to 1977 in West Bengal was caused by a serious inequality in land distribution. Almost half of the population either did not possess any land, or had less than a quarter of a hectare. The landlords constituted less than 7% of the population.

Further, hordes of refugees entered the state during the war in East Pakistan. That caused acute food shortages in the state. The tragic violence of that period forced the communist government to change the policies of land distribution, and those changes brought many people out of the so-called "Naxalite movement" into the mainstream of civil society.

The process of just land distribution is far from done: less than 1% of the cultivable land in the country is fairly distributed. And this unjust land-distribution is only one of the many points in the so-called "unfinished agenda" of the developmental process. I will list other elements of that "unfinished agenda" later, but the point to emphasize here is that the current globalization offers no possibilities to address these social needs.

The Current Globalization. The current globalization started in 1991 during a financial crisis in the country. A new government stabilized the economy, simplified investment in the country, removed many trade barriers, and reformed the tax-system.

On account of this liberalization of the economy, direct foreign investment in the country increased more than 10-fold between 1991 and 1996. Further, in the agricultural sector, this opening up of the economy combined with a series of good monsoons causing rapid agricultural growth from --2.3% in 1991 to 4.9% in 1995.

But this financial blessing evidently happened only for the relatively rich farmers. The fate of the poor did not improve. For rich farmers globalization means, for example, that they have access to new hybrid seeds. Since these hybrids cost much more than the traditional low-yielding varieties, only the rich agriculturists can buy them.

Further, none of the private firms "waste" money in improving the so-called "orphan plants" — grains like sorghum, and legumes like chickpea. Globalization — the opening up of the market — does not mean improvement in the lives of marginal farmers.

The Dangers. There are also other, more general, dangers for agriculture in the current globalization. I will quickly touch on only two, namely, intellectual property rights, and biotechnological considerations.

Intellectual Property Rights in agriculture are supposed to protect both the rights of firms which improve crops, and the rights of agriculturists who possess crop-varieties with valuable traits. In practice, however, there is almost no awareness in rural India about the rights of farmers. Meanwhile, private firms patent crop varieties, as happened recently when researchers in Australia applied for a patent for varieties of chickpea.

They claimed that although they had obtained those varieties from the ICRISAT genebank, they improved the seeds and so have a right to patent them. Only the alarm sounded by a nongovernmental organization woke the institutions of the

CGIAR and the Food and Agriculture Organization (FAO [www.fao.org]) of the United Nations which immediately placed a moratorium on the patenting of these crops, defining them "international public goods".

Biotechnology firms present another aspect of the dangers of the current globalization. These firms invest much in the testing and introduction of new hybrids which both cost a great deal and whose ecological consequences are insufficiently known. Further, these firms protect their investment in ways which are ethically indefensible.

The American biotechnology firm, Monsanto, for example, recently found itself in the center of a political storm on account of its attempt to introduce rice seeds containing the so-called "Terminator" gene; that is, that high-yielding rice hybrid contains a gene that makes sure that seeds from this season's rice crop will not germinate when they are sown the next season: the farmers therefore must again buy seeds from the firm.

Well, once again, a nongovernmental organization sounded the alarm, and a gravely embarrassed Monsanto hastily declared that the firm was only testing the possibilities of the Terminator gene, and had in fact no plans to use it.

In both cases, then, nongovernmental organizations played an important role. On the one hand this emphasizes the role of a free media, and thus, a democratic society (a point which the economist Amartya Sen stresses). But on the other hand, this phenomenon also indicates a disturbing abdication by the state of its duty to protect the most vulnerable sections of the population.

The Heart of the Problem. And that takes us into the heart of the problem with globalization. The state continues to have an important role in a globalizing world — arguably an even more important role than in earlier times when, for example, India was economically impenetrable. But the current economic climate does not encourage one to hope that this realization will be generally accepted. Further, the political instability of coalition governments, and extremely costly "distractions" such as for instance the recent "almost-war" with the neighbor Pakistan, make the task of the state in the developmental process even more difficult.

Meanwhile, one out of three Indians does not have access to safe water; almost one out of two illiterates in the world is an Indian; the largest number of the absolutely poor live in India; the country has more registered unemployed than the total unemployed of the 24 countries of the OECD; and 44 million children in India work.

The list of this "unfinished agenda" continues. Meanwhile, as the 1999 Human Development Report (www.undp.org/hdro/report.html) of the UNDP says: "when research priorities are defined, money, not need, rules — cosmetic drugs and slow-maturing tomatoes appear higher in the priority list than drought-resistant crops and vaccines against malaria."

Only if a democratic government both fulfills its role and forces private firms to participate actively and positively in the process of social development, will globalization not have catastrophic, violent consequences. Only then will be realized the UNDP's aim of "Globalization with a Human Face."

Bibliography

Ausubel, F.M.: *Current Protocols in Molecular Biology*, New York, John Wiley and Sons, 1989.

Brandwein, P.F.: *Sourcebook for the Biological Sciences*, San Diego, Harcourt Brace JOvanovich, 1986.

Brooks, Rodney A.: *Flesh and Machines: How Robots Will Change Us*, New York, Pantheon Books, 2002.

Chrispeels, Maarten : *Plants, Genes and Crop Biotechnology*, Sudbury MA, Jones and Barlett Publishers, 2003.

Cox, Earl D. : *Beyond Humanity: Cyber Evolution and Future Minds*, Roackland, Charles River Media, 1996.

David Sadava: *Plants, Genes and Crop Biotechnology*, Sudbury MA, Jones and Barlett Publishers, 2003.

Dixon, Dougal: *After Man-A Zoology of the Future*, New York, St. Martin's Press, 1981.

Dodds, John H.: *Plant Genetic Engineering*, New York, Cambridge University Press, 1985.

Drexler, K. Eric: *Engines of Creation: The Coming Era of Nanotechnology*, New York, Anchor Books, 1986.

Fransman M, Junne G, Roobeek A: *The Biotechnology Revolution?*, Oxford, Blackwell, 1995.

Friedberg, E.C.: *DNA Repair*, New York, WH Freeman and Company, 1985.

Fumento, Michael: *Bioevolution: How Biotechnology is Changing Our World*, San Francisco, Encounter Books, 2003.

Gaskell G: *Biotechnology-the Making of a Global Controversy*, Cambridge, Cambridge University Press, 2002.

George S. Paul: *Beyond Humanity: Cyber Evolution and Future Minds*, Roackland, Charles River Media, 1996.

Goodman, D.C.: *From Farming to Biotechnology: A Theory of Agro-industrial Development*, Oxford, Blackwell, 1987.

Goodsell, David S.: *Bionanotechnology: Lessons From Nature*, Hoboken, Wiley-Liss, 2004.

Kurzweil, Ray: *The Age of Spiritual Machines*, New York, Penguin Books, 1999.

Moravec, Hans: *Mind Children: The Future of Robot and Human Intelligence*, Cambridge, Harvard University Press, 1988.

Morholt, E. : *Sourcebook for the Biological Sciences*, San Diego, Harcourt Brace JOvanovich, 1986.

Murray, David: *Seeds of Concern: The Genetic Manipulation of Plants*, Sydney, University of New South Wales, 2003.

Old, R.W. : *Principles of Gene Manipulation*, London, Blackwell Scientific Publications, 1989.

Postman, Neil: *Technopoly: The Surrender of Culture to Technology*, New York, Vintage Books, 1992.

Primrose, S.B.: *Principles of Gene Manipulation*, London, Blackwell Scientific Publications, 1989.

Retzer, W.J.: *Biotechnology Workbook*, Englewood Cliffs, Prentice Hall, 1991.

Rifkin, Jeremy: *The Biotech Century*, New York, Penguin Putnam, 1998

Shetty, Kalidas: *Food Biotechnology*, New York, Dekker/CRC Press, 2005.

Slater, R.J.: *Experiments in Molecular Biology*, Clifton, Humana Press, 1986.

Smith, John E.: *Biotechnology*, Cambridge, Cambridge University Press, 2004.

Switzer, R.L.: *Experimental Biochemistry*, New York, W.H. Freeman and Company, 1977.

Towle, Albert: *Modern Biology*, Austin, Holt, Rinehart and Winston, 1988.

Walden, Richard: *Genetic Transformation in Plants*, England, Open University Press, 1988.

Winston, Mark L.: *Travels in the Genetically Modified Zone*, Cambridge, Harvard University, 2004.

INDEX

■■■